SOIL MOISTURE

Edited by **Gabriela Civeira**

Soil Moisture
http://dx.doi.org/10.5772/intechopen.73742
Edited by Gabriela Civeira

Contributors

Dai Matsushima, Judith Ramos, Jesús Gracia Sánchez, Tania Rodríguez-Martínez, José Zúñiga, Hongquan Wang, Lucieta G. Martorano, Araya A. Berhe, José Reinaldo Da Silva Cabral De Moraes, Douglas Cavalcante Costa, Aline Michelle Da Silva Barbosa, Marcelo Coelho Coelho Marques, Ayllan Rayanne Da Silva Lima, William Chavarriaga, Gabriela Civeira

First published in London, United Kingdom, 2019 by IntechOpen
IntechOpen is the global imprint of INTECHOPEN LIMITED, registered in England and Wales, registration number: 11086078, The Shard, 25th floor, 32 London Bridge Street
London, SE19SG – United Kingdom
Printed in Croatia

British Library Cataloguing-in-Publication Data
A catalogue record for this book is available from the British Library

Additional hard copies can be obtained from orders@intechopen.com

Soil Moisture, Edited by Gabriela Civeira
p. cm.
Print ISBN 978-1-78985-103-8
Online ISBN 978-1-78985-104-5

For EU product safety concerns:
IN TECH d.o.o., Prolaz Marije Krucifikse Kozulić 3, 51000 Rijeka, Croatia,
info@intechopen.com or visit our website at intechopen.com.

Meet the editor

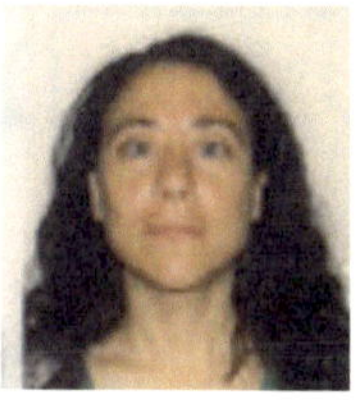

Dr. Gabriela Civeira is currently an adjunct professor of agrosystems environmental impact at the Faculty of Agronomy, Morón University, Buenos Aires and was previously an adjunct professor of ecology at the same university. Born in Buenos Aires, Argentina, Gabriela was educated at Westminster School in Banfield, Argentina and graduated with a Bachelor's degree in agronomy engineering, earned a scholarship in order to pursue research and teaching abilities at the Faculty of Agronomy from the University of Buenos Aires. Following her university work from research activities, Gabriela then attended University Graduate School, University of Buenos Aires gaining her Masters in soil science degree before embarking on an academic career, which has both embraced teaching undergraduate and postgraduate students and research. During 2016, Gabriela earned her Doctoral degree at the University of La Coruña with a thesis on ecosystem services in urban environments and landscape planning.

Contents

Preface

Soil moisture has been the main research subject of soil physics, and chemical and biological properties. In addition to the major role it plays in the agricultural system, soil moisture has been one of the primary tests often done in research and production studies that provides important contributions to our knowledge of soil biogeochemistry, microorganisms activity, organic matter content and crop yield. For example, soil genesis and development were derived from pedological processes that only occur if soil has adequate water content. Nowadays, the use of soil moisture for research purposes is increasing, and the variety of tests performed on water content are also diversifying each day. It is clear that soil moisture will continue to serve as experimental and food production material in the most important areas of agricultural science in the near future. This is because the water found and used by humans is in groundwater bodies. These water bodies are replenished by drainage through the soil, and then the rate of groundwater recharge is greatest when rainfall inputs to the soil exceed evapotranspiration losses. Knowledge of the recharge features of catchments is indispensable to water supply management, planning, flood prediction and floodplain regulation. Even though it is not possible to follow the increasing body of knowledge in this area, this book aims to provide the reader with a wide-ranging summary of the diverse types of studies in which soil moisture is the focal research object, increasing information and understanding of the current knowledge that have been utilized in the area of soil water content.

Dr. Gabriela Civeira
Instituto Nacional de Tecnologia Agropecuaria
Villa Udaondo, Hurlingham, Argentina

Section 1

Introductory Chapter

Chapter 1

Introductory Chapter: Soil Moisture

Gabriela Civeira

Additional information is available at the end of the chapter

http://dx.doi.org/10.5772/intechopen.83603

1. Introduction

Water is the common medium for several life processes. There is no lifecycle without water. Moreover, in plants situation, the uptake of nutrients through the roots is intermediated by soil water. Consequently, water and soil are the elementary requirements for the life and growth of plants. Nevertheless, water availability in soils is restricted, and therefore, its administration and measurement turn into an indispensable issue [1]. Water management includes understanding about the release pattern and storage of water by the soil. This knowledge on storage and release also needs understanding about physical biological and chemical properties and soil processes [2]. Soil properties strongly affect processes, which take place on it: among others, soil remains warm or cool, dry or wet, aerobic or anaerobic, friable or rigid, highly porous or compact, aggregated or dispersed, impermeable or permeable, eroded or preserved, saline or unsalted, nutrient-rich or percolated, etc. The last regulates whether the soil water content can be a good or bad medium for various varieties of plants as well as for different methods of production and whether it can conveniently help as an operative convertor for environmental contaminants, rather than as a transmitter of them. The soil moisture content correspondingly referred to as water content and is an indicator of the quantity of water existing in soil. By means, moisture content in soils is the relation of water quantity in a portion to the quantity of solids in the soil sample, expressed as a proportion for example (percentage) [3].

Related to other constituents of the hydrological cycle, the dimension of soil moisture is slight; nevertheless, it is of essential significance to numerous biological, hydrological, and biogeochemical processes. Soil moisture data are appreciated to an extensive variety of administration agencies and private corporations concerned through meteorological conditions and climate, inundation control and runoff potential, soil erosion, water reservoir controlling, and water quality, among other subjects. Soil moisture is a crucial variable in governing the exchange of water and warmth energy among the land surface and the atmosphere through plant transpiration and soil evaporation [3, 4]. As a consequence, soil moisture shows a vital protagonist in the development of climate patterns and the creation of precipitation. Climate prediction models have shown that enhanced description of superficial soil moisture and plants can lead to noteworthy estimation enhancements. Likewise, soil moisture intensely disturbs the volume of precipitation that runs off inside proximate rivers and streams. Worldwide, significant landscapes with dry or wet shallow areas have been observed to report positive response on succeeding rainfall patterns. Also, soil moisture data can be used for water reservoir content and managing, early advice of deficiencies, irrigation planning, and crop yield estimation [1, 5].

Even though the importance of soil moisture data availability, widespread and constant quantifications of soil moisture is negligible. As mentioned before, there is still a deficiency of conclusive methods of soil moisture measurement worldwide [2, 4]. Undoubtedly, a necessity subsists for constant measurements of shallow soil moisture through a worldwide reportage. These book chapters aim to understand soil moisture behavior, including classical and new tools, which cover a wide range of moisture indices. Classical techniques of water content analysis showed in this book included soil moisture measurements obtained by drying a known quantity of soil sample in an oven and pore spaces percentage of the soil volume not occupied by solids and assumedly occupied by air and water [4]. Among new water content estimation tools, this book included remote sensing and TDR technique approximations, which use the energy emitted from the soil layers [2]. Finally, this book was designed to meet the needs of researchers, students, and academics and pursued the subsequent objectives: (a) assess the influence of soil practices on soil moisture release and retention properties, (b) relate the water indices with the soil moisture properties and characteristics, (c) estimate soil moisture release and retention among different soil types, and (d) evaluate and develop suitable models to estimate soil moisture content at different scales and levels of analysis. In order to meet these objectives, scientists and academics from different continents have been involved in the field and laboratory experiments to address some answers, which dare to face with emerging technology in soil moisture assessment.

Author details

Gabriela Civeira

Address all correspondence to: gciveira@agro.uba.ar

Instituto Nacional de Tecnologia Agropecuaria, Hurlingham, Argentina

References

[1] Briggs LJ. The Mechanics of Soil Moisture Paperback. Wentworth Press; 2016. 38 p. ISBN-10: 1373429763, ISBN-13: 978-1373429766

[2] Zwartendijk BW, Van Meerveld HJ, Ghimire CP, Bruijnzeel LA, Ravelona M, Jones JPG. Rebuilding soil hydrological functioning after swidden agriculture in eastern Madagascar. Agriculture, Ecosystems & Environment. 2017;**239**:101-111

[3] Lin H, Bouma J, Wilding LP, Richardson JL, Kutilek M, Nielsen DR. Advances in hydropedology. Advances in Agronomy. 2005;**85**:1-89. ISSN: 0065-2113

[4] Richards LA. Methods of measuring soil moisture tension. Soil Science. 1949;**68**(1):95

[5] Lin H, Bouma J, Pachepsky Y. Hydropedology: Bridging disciplines, scales and data. Geoderma. 2006;**131**(3-4):255-406

Section 2

Estimation Techniques

Chapter 2

Thermal Inertia-Based Method for Estimating Soil Moisture

Dai Matsushima

Additional information is available at the end of the chapter

http://dx.doi.org/10.5772/intechopen.80252

Abstract

Thermal inertia is a parameter that characterizes a property of soil that is defined as the square root of the product of the volumetric heat capacity and thermal conductivity. Both properties increase as soil moisture increases. Therefore, soil moisture can be inversely determined using thermal inertia if a relationship between the parameters is obtained in advance. In this chapter, methods for estimating surface soil moisture using thermal inertia are comprehensively reviewed, with emphases on the followings: How thermal inertia is retrieved accurately from a surface heat balance model, and how it is accurately converted to surface soil moisture. In addition, the advantages and disadvantages of the thermal inertia methods are discussed and compared to microwave-based methods, such as spatial resolution and the sky conditions. Precise and accurate data from earth observing satellites are indispensable for estimating the spatial distribution of thermal inertia at a high resolution. On the other hand, data assimilation methods are rapidly developing, which may be competitive with thermal inertia methods. Finally, applications of thermal inertia methods are described and discussed for future explorations, such as dust emission in relation to soil moisture, and estimating regional water budgets by combining other satellite data.

Keywords: thermal inertia, soil moisture, thermal-infrared band, land surface temperature, earth observing satellite, surface heat balance, force-restore model, Fourier series expansion, microwave-based method

1. Introduction: thermal inertia of unsaturated soil

Thermal inertia P is one of the parameters used to characterize the thermal properties of soil and is defined as the square root of the product of the volumetric heat capacity C and thermal conductivity λ, which is given as

$$P = \sqrt{C\lambda}. \quad (1)$$

Thermal inertia appears in the formulation of the ground heat flux when it is formulated with a unique variable known as the land surface temperature (LST), and one does not have to consider the vertical profile of the soil temperature.

In terms of the relation between soil thermal properties and soil moisture, both the volumetric heat capacity C, which is the product of specific heat c and the bulk density ρ of the soil, and the thermal conductivity λ increase as the soil moisture increases. Accordingly, thermal inertia P also increases as soil moisture increases. Therefore, soil moisture can be estimated inversely if the thermal inertia value is known (**Figure 1**). The volumetric heat capacity is a moderate linear function of soil moisture [1]. By contrast, thermal conductivity has a strong nonlinearity with soil moisture, making it difficult to parameterize thermal conductivity and hence thermal inertia.

Thermal inertia is effective when the time series of the surface temperature is available, but the vertical profile of the soil temperature is not available. The differential equation for heat diffusion is given as

$$\frac{\partial T(z,t)}{\partial t} = \frac{\lambda}{C}\frac{\partial^2 T(z,t)}{\partial z^2}, \quad (2)$$

where $T(z,t)$ is the soil temperature at depth z and time t, which is the difference from a constant value at an infinite depth. This equation is solved to reproduce the daily and yearly

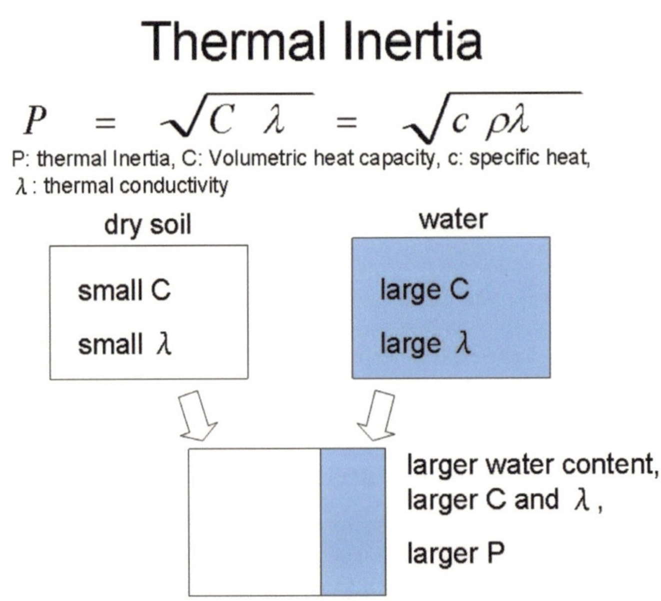

Figure 1. Schematic of the relationship between volumetric heat capacity, thermal conductivity, and thermal inertia in terms of soil moisture.

periodic cycles of soil temperature, under the boundary conditions that the temperature change is sinusoidal at the surface and constant at an infinite depth. The basic solution is as follows using the complex number expression:

$$T(z,t) = A\exp\left(-\frac{z}{d}\right)\cdot \exp\left[i\left(\omega t - \frac{z}{d}\right)\right], \tag{3}$$

if the surface boundary condition is $T(0,t) = A\cdot \exp[i\omega t]$. In Eq. (3), A is the amplitude of a periodic change with an angular velocity ω, d is the scale depth at which the amplitude is e^{-1} of the surface value, and i is the imaginary number $\sqrt{-1}$. The scale depth d is formulated as

$$d = \sqrt{\frac{2\lambda}{\omega C}}. \tag{4}$$

Eq. (3) is rewritten as follows using the real number expression:

$$T(z,t) = A\exp\left(-\frac{z}{d}\right)\cdot \cos\left[\omega t - \frac{z}{d}\right], \tag{5}$$

if the surface boundary condition is $T(0,t) = A\cdot \cos\omega t$.

The conductive heat flux in soil at depth z and time t, $G(z,t)$, is defined as

$$G(z,t) = -\lambda\frac{\partial T(z,t)}{\partial z}, \tag{6}$$

where the vertical profile of soil temperature is obviously required to calculate the soil heat flux. However, when the soil temperature solution (3) or (5) is applied to the soil heat flux Eq. (6) and then z is set to zero, it only uses the time series of surface temperature as follows:

$$G(0,t) = \sqrt{\frac{C\lambda}{2\omega}}\left[\frac{\partial T(0,t)}{\partial t} + \omega T(0,t)\right]. \tag{7}$$

In deriving Eq. (7), the following relation derived from Eq. (3),

$$\frac{\partial T(z,t)}{\partial t} = i\omega T(z,t), \tag{8}$$

is used. Eq. (7) is known as the force-restore method (FRM) for calculating the surface soil heat flux [2]. In Eq. (7), the numerator of the parameter is defined as thermal inertia P, which is given as

$$P = \sqrt{C\lambda} = \sqrt{c\rho\lambda}. \tag{9}$$

The above derivation describes thermal inertia being effective for quantifying soil thermal properties when only the LST is known, which leads to the analysis of the land surface processes using satellite LSTs. The above procedure leads to the studies proposed by Matsushima and co-researchers, which are described in Section 2.

Considerable effort has been made to estimate the thermal inertia of the Earth's surface mainly using LST data from satellites. Most of this effort has been concentrated on retrieving daily values of thermal inertia due to the availability of daily maximum and minimum LSTs observed from polar orbiting or geostationary satellites. Models using these types of satellite LSTs are based on the Earth's surface energy balance principle, which includes not only the radiation budget but also turbulent heat flux. A Fourier series expansion was introduced to solve Eq. (1) under the above Earth's surface boundary conditions using the solution of the real number expression, Eq. (5). Models have been improved from those using only the two daily extreme LSTs [3–8] to those using LSTs that are irrespective of time in a diurnal change [9, 10] and other significant studies that follow a series of important proposals by Xue and Cracknell [11–14], which are also described in Section 2 when compared with studies by Matsushima and co-researchers.

This chapter reviews former and state-of-the-art methods for estimating soil moisture by exploring the relationship between thermal inertia and soil moisture. Section 2 reviews past developments of methods for thermal inertia retrieval from land surface models. Section 3 describes how thermal inertia is experimentally observed, and how it is retrieved from land surface models in terms of the Xue and Cracknell-based models and the Matsushima models. Section 4 describes several semi-empirical parameterizations of thermal inertia in terms of soil moisture. Section 5 describes applications of thermal inertia for analyzing hydrometeorological phenomena around the Earth's surface and also discusses further exploration of thermal inertia itself and its applications. Section 6 presents conclusions.

2. Thermal inertia retrieval from energy balance models

Thermal inertia retrieval from an energy balance model of the Earth's surface began with a geological context in which different rocks or minerals respond differently to the incident solar radiation. Then, researchers' interests moved to the thermal inertia change according to soil moisture, which was coincident with the use of data obtained by sun-synchronous polar orbiting satellites that gave diurnal cycle of LSTs. Retrieval thermal inertia solely according to the soil moisture of the Earth's surface was developed during the last five decades. Most of the proposed methods employed the Fourier series of the LST diurnal variation, which was incorporated in a heat balance model of the Earth's surface.

A comprehensive model for retrieving thermal inertia using the energy balance model for the Earth's surface boundary conditions and the Fourier series for the diurnal change of the surface temperature was proposed by Price [3–5]. Price [6] made the terms of the turbulent heat flux (sensible and latent heat) simpler than previous studies to focus on retrieving thermal inertia. These studies used satellite LST measurements twice a day as the daily maximum and minimum LSTs, corresponding to daytime and nighttime, respectively, and substituted the LSTs into the first component (24-h period) of the Fourier series to calculate thermal inertia. Namely, they approximated the time-differential term of Eq. (7) as $\frac{\partial T(0,t)}{\partial t} \to \frac{\Delta T(0,t)}{\Delta t}$, where $\Delta T(0,t)$ is the difference between daily maximum and minimum LSTs, and Δt is the time difference of the two measurements.

Based on a series of studies performed by Price [3–6], Xue and Cracknell [11–14] proposed improved methods, which showed that data from satellites were good enough to accurately retrieve thermal inertia as well as using the time of the maximum LST. These models used the first- and second-order harmonics of the diurnal change (24- and 12-h periods) to fit the LST change considering the phase differences of both components to insolation. Thermal inertia was obtained from analytical but relatively complicated formulations. Based on the series of models proposed by Xue and Cracknell (hereinafter the XC model), several improved methods were proposed in terms of the timing of satellite measurements, actual timing of the diurnal maximum and minimum LSTs, and difference in LST change between daytime and nighttime. The details of the above schemes are described in Section 3.2.

Other than the above methods, Matsushima [15] applied the FRM to the surface heat balance model to retrieve thermal inertia. The FRM is also based on a sinusoidal boundary condition at the surface and the heat diffusion equation, which is essentially the same as the models based on the XC model. The Matsushima model [15] employed an FRM that was designed not only to mostly respond to the diurnal change but also to more rapid changes according to the temporal resolution of input the variables (insolation, air temperature, etc.). A change of the LST over a period of approximately a few hours was fairly reproduced by the FRM that had a characteristic period of 24 h, as illustrated in [16]. Similar results were found in other studies [2, 17], or higher-frequency nonsinusoidal forcing did not significantly affect the LST prediction [18]. This means that the FRM can reproduce temporal changes that have a wide range of LST frequencies via its relatively simple formulation. Using this method, the timing of satellite LST measurements was arbitrary in principle, irrespective of the daily maximum and minimum, but was more accurate for thermal inertia retrieval that the LSTs measured both in the daytime and in the nighttime, as shown in [19]. The accuracy of thermal inertia retrieval is improved if the coefficients of the atmospheric turbulent heat flux are set differently in the daytime and nighttime, as illustrated in [16]. The details are described in Section 3.2 when compared with the XC model.

3. Thermal inertia retrieval according to the spatial and temporal resolutions

3.1. A simple method using in-situ field measurements on a local scale

A simple method for estimating thermal inertia using simply measured surface radiative temperatures can be performed based on a finite difference form of Eq. (7), which is given as

$$\overline{G(0)} = \sqrt{\frac{P}{2\omega}}\left[\frac{\Delta T(0)}{\Delta t} + \omega\overline{T(0)}\right], \tag{10}$$

where $\Delta T(0)$ is a significant increase in LST during a relatively short-time span Δt (e.g., 30 min), $\overline{T(0)}$ is the temporal average of the LST difference from the soil temperature at infinite depths during the time span (practically, the daily average surface or air temperatures can be used instead of the temperature at an infinite depth), and $\overline{G(0)}$ is soil heat flux at the surface averaged over the time span Δt. Matsushima et al. [20] estimated the thermal inertia of asphalt pavement based on Eq. (10) using data from $T(0)$ measured by a portable radiative thermometer approximately 1 m above the surface as well as a copper-constantan thermal couple on the surface and $\overline{G(0)}$ measured by a heat flux plate on the asphalt surface over 30 min under a summer daytime clear sky, assuming that the daily average thermocouple temperature was the temperature at an infinite depth. The results of the thermal inertia were 1140 J m^{-2} K^{-1} $s^{-1/2}$ and 1350 m^{-2} K^{-1} $s^{-1/2}$ using the LSTs measured by the portable radiative thermometer and the thermocouple, respectively. The values agree well with the standard value of 1220 m^{-2} K^{-1} $s^{-1/2}$ in the literature [21] (actually, this value is based on the volumetric heat capacity and thermal conductivity, calculated using the definition of thermal inertia Eq. (1)). The above result shows that thermal inertia can be estimated at a local scale using simple measurement equipment and also shows the feasibility of thermal inertia estimation using airborne and satellite thermal-infrared thermometry.

3.2. Model methods using multiple satellite data

Earth observing satellites are divided into polar-orbiting and geostationary. Both types have thermal-infrared bands to estimate LST. The temporal resolution of the geostationary satellites is superior to that of the polar-orbiting ones. By contrast, the spatial resolution and temperature accuracy of the geostationary satellites have been improving but are not yet superior to those of the polar-orbiting one [22, 23]. Therefore, in this section, models using polar-orbiting satellites are described and discussed.

Most of the models proposed so far for estimating thermal inertia were based on the model proposed by Xue and Cracknell [13] (XC model). This model is based on solving the thermal diffusion equation using the Fourier series expansion, which is described in Section 1. This model uses the first 24-h period and the second 12-h period Fourier harmonics of the sinusoidal components to reproduce the diurnal variation of LSTs. However, the two components were not always enough to reproduce an actual diurnal LST change. In these cases, the phase differences from the diurnal change of the insolation of respective components have to be adjusted. Measured values in thermal-infrared bands were used to calculate the LSTs. The LSTs are almost the daily maximum and minimum, which are suitable for accurately estimating thermal inertia. Various improvements were proposed based on the XC model. Among these improvements, a method using four satellite LSTs during a diurnal cycle irrespective of the daily maximum and minimum successfully retrieved thermal inertia [9]. Schemes for the phase and amplitude adjustments were also introduced. The phase adjustment was applied to the time-difference between the actual timing of the maximum and minimum LSTs and the overpass timings of the satellite measurements. This adjustment scheme was to have the satellite overpass timing approach the daily maximum or minimum LSTs using a cosine function for the phase difference when their estimated times were given [24]. On the other hand, an adjustment that decayed the LST amplitude after sunset was applied during nighttime due to the LST change being smaller than that in the daytime to avoid overestimating nighttime cooling [25]. These adjustments allowed arbitrary and many timings of the satellite

LSTs to be incorporated into the models for thermal inertia retrieval. The formulations for estimating thermal inertia were analytical but relatively complicated to describe and grasp. Moreover, various approximations were adopted to avoid the implicit formulations that required iterative calculations to retrieve thermal inertia. The input data required are according to individual models but are almost limited to the satellite LSTs, and some parameters with regard to the insolation are also required to specialize in retrieving thermal inertia from other parameters regarding the land surface processes. Models that simply address turbulent heat flux were proposed assuming that turbulent heat flux is proportional to the temperature difference between surface and atmosphere [26], or was ingeniously neglected [27]. Maltese et al. [10] proposed an XC-based model that only used the first Fourier component, which was enough to accurately retrieve thermal inertia using three LSTs during a diurnal cycle.

In contrast to the XC-based models, a series of models proposed by Matsushima and co-researchers (the most recent one is [16], hereafter referred to as M2018) was essentially based on the sinusoidal solution of the thermal diffusion equation. However, they adopted the FRM to avoid complicated analytical formulations for estimating thermal inertia while maintaining linearity. The FRM requires time integration to calculate the surface temperature as shown in Eq. (7). The characteristic frequency ω was set to the diurnal change (= $2\pi/86400(s^{-1})$) in M2018. Hence, the FRM appeared to be able to only reproduce the sinusoidal changes whose frequencies were near the characteristic frequency. However, shorter period changes were reproduced according to the input data changes shown in Figure 4 of M2018 because the force term (the time derivative term) was more temporally sensitive than the restore term (the product of the frequency and temperature difference from the daily average). The boundary conditions at the surface, the left side of Eq. (7), are also required. The boundary conditions are converted to the budget of net radiation, and sensible and latent heat at the surface using the heat balance equation. This requirement for the surface boundary conditions is as same as that of the XC-based models. M2018 was not specialized at retrieving thermal inertia, but other parameters with regard to the sensible and the latent heat flux, including the diurnal time series of insolation, air temperature, specific humidity, and wind speed, were required as input data. Also, the surface albedo and leaf area index were used as parameters, and the LSTs for the model optimization are described below. The above types of data were required, but all were readily available from satellite and meteorological data archives through the Internet, that is, a special observation was not needed. Instead of requiring many types of input data, the M2018 formulation was relatively simple and did not require phase and amplitude adjustments. The shift in values of the bulk transfer coefficients for the sensible and latent heat flux in the daytime and nighttime was required only to improve the accuracy of thermal inertia retrieval, which was approximately equivalent to the amplitude adjustment in the XC-based model by Schädlich et al. [25]. The time integration did not require an implicit scheme, but an optimization algorithm was required to retrieve thermal inertia and the other parameters at the same time. The optimization algorithm (the downhill simplex method that was employed in M2018) took time to retrieve parameters; hence, the M2018 had no advantage for the world-wide spatial scale and the temporal scale of several decades. In M2018, daily values of thermal inertia-derived soil moisture were estimated with a 3-km spatial resolution at $2° \times 2°$ in latitude and longitude using the 1-km spatial resolution of Moderate Resolution Imaging

Items	XC-based models	M2018 [16]
Basic equation	Differential equation of heat diffusion	
Boundary conditions	Sinusoidal function at the Earth's surface constant at infinite depth	
Model	Fourier series expansion	Two-source energy balance (TSEB) model based on force-restore model (FRM)
Solution	Fourier series components (first component only [10] or both first and second components [11–14] according to models)	Time integration of the respective surface temperatures of the two sources
Input data and parameters	LSTs parameters in terms of insolation parameters in terms of turbulent heat flux (according to the models)	LSTs insolation air temperature, specific humidity, wind speed albedo, and the leaf area index of the surface
Thermal inertia retrieval	Analytical solution	Model optimization (with other parameters regarding turbulent heat)
Number of diurnal LST measurements	Two (without timing adjustment for maximum and minimum [11–14]); two (with timing adjustment [24]); three or more [9, 10]	Arbitrarily determined (at least two—one in the daytime and the other in the nighttime—are suitable) [19]
Phase adjustment of LST	Needed in most models	Adjusted through time integration incorporating input time series

Table 1. Comparison of the differences between the XC-based models and M2018 model.

Spectroradiometer (MODIS) LST (MOD11_L2 and MYD11_L2). Thermal inertia retrieval at a 2-km resolution was, therefore, possible in principle [28].

Comparisons of the differences between the XC-based models and M2018 model are provided in **Table 1**.

4. How soil moisture is derived from thermal inertia

4.1. Combination of $C - \theta$ and $\lambda - \theta$ relations

One of the principal methods for deriving the relationship between thermal inertia P and soil moisture (volumetric water content in most cases) θ uses the definition of thermal inertia $P = \sqrt{c\rho\lambda} = \sqrt{C\lambda}$. Specifically, the effective models for using soil moisture to determine the volumetric heat capacity and thermal conductivity, respectively, proposed by de Vries [29] were used in several thermal inertia models [7, 8]. The volumetric heat capacity is formulated as

$$C = C_w\theta + C_m(1 - \theta_*), \tag{11}$$

where C_w and C_m are the volumetric heat capacities of water and minerals, respectively, and θ_* is the soil moisture at saturation, in other words, the porosity of the soil, and the thermal conductivity λ is formulated as

$$\lambda = \sum_{i=0}^{N} K_i X_i \lambda_i / \sum_{i=0}^{N} K_i X_i, \tag{12}$$

where N is the number of types of granules and particles that make up the soil, including the minerals, organic matter, water, and air inside the soil. Each component has a thermal conductivity λ_i and a volume fraction X_i, and K_i is a weighting factor that is the ratio of the average temperature gradient in the granules of the i-th component to the average temperature gradient in the medium. See Appendix for details on the formulation of K_i. Minacapilli et al. [30] combined the linear relation of the volumetric heat capacity proposed by de Vries [29] and an empirical parameterization for thermal conductivity proposed by Lu et al. [31] to derive thermal inertia. The proposed models for the thermal conductivity of unsaturated soil have been expanded, and the details are provided in a review by Dong et al. [32].

Ma and Xue [33] proposed an empirical parameterization that often appears in the literature. This parameterization calculates thermal inertia for a given soil moisture (gravimetric soil water content) when the soil (mineral) density and water density are known.

Noilhan and Planton [34] derived the relationship between thermal inertia and soil moisture in another way. This method was basically a combination of thermal inertia, the relationship between soil moisture and matric potential of soil, and a parameterization of the thermal conductivity as a function of the matric potential proposed by McCumber and Pielke [35]. In their paper, they showed the relationship between soil moisture and soil thermal coefficient C_s, which was defined in their paper, but was able to be rearranged to according to the relationship between M2018 thermal inertia and soil moisture, which is formulated as

$$C_s = \frac{2}{P\sqrt{\tau/\pi}}, \tag{13}$$

and

$$C_s = C_{s,*}\left[\frac{\theta_*}{\max(\theta, \theta_w)}\right]^{(2\ln 10/b)}, \tag{14}$$

where $\tau = 2\pi/\omega$, and subscripts * and w denote the saturation and wilting points, respectively. Substituting Eq. (13) into Eq. (14), after several calculations, yields

$$\theta = sP^{(2\ln 10/b)} \tag{15}$$

where

$$s = \theta_*\left(\frac{C_{s,*}}{2}\sqrt{\frac{\tau}{\pi}}\right)^{(2\ln 10/b)} \tag{16}$$

The constant s depends on the parameters b, θ_*, and $C_{s,*}$, of the Clapp and Hornberger parameterization [36]. In particular, parameter b is related to 11 categories of soil types determined by the United States Department of Agriculture (USDA), and parameter b is a predictive parameter of the clay ratio in soil [37].

4.2. Analogous to Johansen's thermal conductivity model

Johansen [38] proposed a model for determining thermal conductivity as a function of soil moisture. The concept of the model is that thermal conductivity is formulated as a universal function of soil moisture and that the function shape is determined by parameters of the formulation. The parameters are determined according to the soil type, such as sand, loam, silt, and clay. The generalized form of the parameterization is given as

$$\lambda = \lambda_{dry} + \left(\lambda_{*} - \lambda_{dry}\right)K_p \tag{17}$$

where the subscript *dry* denotes zero soil moisture, and K_p is the universal Kersten function. The formulation is defined as a function of soil moisture from zero to the saturation point (porosity). Then, the problem is reduced to determine the specific formulation and its parameter values. The specific form of the Kersten function is a power function, and the curve shape depends on the power according to the soil type, which has strong nonlinearity in most cases.

Murray and Verhoef [39] applied the above Johansen type model to thermal inertia parameterization as follows:

$$P = P_{dry} + \left(P_{*} - P_{dry}\right)K_p \tag{18}$$

To calculate the thermal inertia value, parameters P_{dry} and P_{*} have to be determined, and the formulation of K_p is given as the parameterization proposed by Lu et al. [31] in [39], as

$$K_p = \exp\left[\gamma\left(1 - S_r^{\gamma-\delta}\right)\right], \tag{19}$$

where γ and δ are the coefficients for optimization according to the soil type, and S_r is the soil moisture normalized by saturation. Lu et al. [40] proposed a similar parameterization as that in [39]. Minacapilli et al. [41] tested the performance of the Murray and Verhoef model [39] and extended the Johansen model concept to the apparent thermal inertia. Recently, Lu et al. [42] showed that P_{dry} was parameterized as a function of porosity, in other words, the soil clay ratio, improving the accuracy of thermal inertia retrieval.

Again, thermal inertia is the square root of the product of volumetric heat capacity and thermal conductivity, and volumetric heat capacity increases modestly according to soil moisture. By contrast, thermal conductivity has strong nonlinearity compared to soil moisture. Therefore, thermal inertia can be formulated as a nonlinear function, and even, the square root operates the product $C\lambda$. Lu et al. [40] applied the formulation to thermal inertia and determined the parameter values according to three soil types. The minimum and maximum thermal inertia values range over soil moisture from zero to saturation. If some amount of error is added to the retrieved value of thermal inertia from a model calculation or laboratory experiment, the value may be less than the minimum, and hence, soil moisture cannot be calculated due to K_p being negative.

5. Applications and discussion for future exploration

5.1. Advantages and disadvantages compared to microwave-based methods

Thermal inertia-derived soil moisture can be estimated by combining methods as described in Sections 3 and 4. An advantage of the thermal inertia method that uses satellite data is that the spatial resolution is a couple of kilometers, which is much more precise than that of the microwave-based method, which has the spatial resolution of several tens of kilometers. However, there are also disadvantages, such as the precision and accuracy of thermal inertia retrieval being affected by the sky conditions, especially clouds, which are the weakest point in using the thermal-infrared bands. A recent study showed that the microwave brightness temperatures complemented the thermal-infrared derived LST, but instead of this, the spatial resolution of the thermal-infrared LST had to be sacrificed [43]. Another disadvantage is that the thermal inertia of a surface covered with dense vegetation is difficult to retrieve. Soil moisture retrieval using the microwave bands also has the same problem. Thermal inertia retrieval over a surface covered with sparse vegetation has been achieved in many studies in which M2018 is categorized in the two-source energy balance (TSEB) concept [44, 45]. In M2018, the vegetation canopy is modeled according to its surface temperature, the three parameters that should be optimized, and the leaf area index, which is given as satellite data. The effectiveness of the TSEB model is not only to retrieve thermal inertia but also possibly to accurately calculate heat flux with regard to the surface heat balance. The denser the vegetation, the less accurate the thermal inertia retrieval. It should be noted that the thermal inertia-derived soil moisture is calculated through a simple TSEB model as well as the surface heat flux, including evapotranspiration.

According to the above advantages and disadvantages, soil moisture derivation for a surface is more effective in arid and semi-arid regions where clear sky conditions overwhelm other conditions and where there are spatial soil moisture contrasts, for example, between an oasis and other land cover, as well as significant temporal changes, for example, from just after to approximately 1 week after a rainfall.

The optimization scheme for the thermal inertia retrieval is crucial to save calculation time. M2018 uses the downhill simplex method [46], which is generally suitable for optimizing less than approximately five parameters. It takes approximately 20–40 s to retrieve seven parameters including the thermal inertia of one grid in M2018 using a workstation. The downhill simplex method has the advantage of not diverging in the optimizing process, but the algorithm is not simple and requires a long time to complete. A more efficient optimization scheme needs to be explored.

5.2. Assimilation with microwave-based data

Data assimilation procedures are downscaled schemes of microwave-based soil moisture, which has a scale of several tens of kilometers, to one to a couple of kilometers. These schemes have recently been improved [47–50] using visible, near-infrared, and thermal-infrared satellite data, which have more precise spatial resolution than microwaves. These procedures can be competitive with thermal inertia procedures to derive surface soil moisture. However, one weak point with regard to microwave-based soil moisture (soil moisture active passive: SMAP) was noted, and it was a dry down process occurred after an antecedent rainfall that was too rapid for in-situ soil moisture measurement [50]. By contrast, the thermal inertia-derived soil moisture agreed fairly well with the in-situ soil moisture found in several dry down processes (M2018). This agreement may be because the sensing depth of the surface microwave-based soil moisture was shallower than the in-situ measurement and sensitive to the soil moisture itself [51], whereas the representative depth scale of the FRM is not as sensitive to soil moisture and almost agrees with the in-situ measuring depth (M2018). Regarding the spatial resolution of the satellite sensors, an Earth observing satellite with a more precise spatial resolution in the visible, near-infrared, and thermal-infrared bands, the Global Change Observation Mission-Climate (GCOM-C), was recently launched in 2017 by the Japan Aerospace Exploration Agency (JAXA), and the data will be available for general use within 1 year. Its LST spatial resolution is 500 m, twice than that of the MODIS resolution, which will benefit both data assimilation and thermal inertia procedures. Another GCOM-C type satellite will hopefully be able to be operated like MODIS. On the other hand, microwave-based soil moisture can be obtained almost every day regardless of the sky conditions (leading to partial lack of data in some regions due to the satellite orbit). There are trade-offs that have between the above described advantages and disadvantages of the respective procedures.

The Global Satellite Mapping of Precipitation (GSMaP) [52] operated by JAXA is a system that measures the spatio-temporal distribution of precipitation at the Earth's surface on a 0.1°-spatial scale and a 1-hour temporal scale, and the latest data are added every hour. In arid and semi-arid regions far from rivers, short-term discharge and infiltration should be negligible, accordingly the water budget is calculated using the thermal inertia-derived soil moisture and GSMaP precipitation. Currently, there is not adequate accuracy for both variables to calculate a water budget, but it is worth tackling this issue to estimate regional water cycles and resources.

5.3. Dust emission

Dust emissions in arid and semi-arid regions have posed serious problems such as soil nutrition loss, crop and vegetation damage, and air quality deterioration. Dust emissions, for example, from Northeast Asia, influence not only individual arid or semi-arid regions but also regions across national boundaries and seas because some dust is raised by strong atmospheric convection and carried by strong westerly winds [53]. Dust emission from the Saharan Desert often harms the surrounding regions including regions far from Africa [53]. Wind erosion from agricultural land often causes local and regional problems according to tillage practices [54]. To predict these dust emissions in advance, monitoring and prediction of the surface soil moisture distribution over an area where dust emission occurrences are concentrated are important. Scheidt et al. [24] examined the spatial distribution of thermal inertia for five soil types in a desert of approximately 20 km using eight couples of the daytime and nighttime thermal-infrared surface temperatures observed by the Advanced Spaceborne Thermal Emission and Reflection (ASTER) and MODIS and then estimated the threshold wind speed based on the estimated thermal inertia values after Fécan [55]. The threshold wind speed or friction velocity for dust emissions according to surface soil moisture was examined using carefully designed wind tunnel experiments with multiple soil types, and the threshold friction velocity was found to be related to the soil matric potential and not aligned with the gravimetric soil moisture for the examined soil types [55–58]. The matric potential is not as readily available as the soil moisture, otherwise the function connecting the two variables is known in advance. Considering that it should be difficult to obtain the relationship between thermal inertia and matric potential, and practical relationships between thermal inertia-derived soil moisture and threshold wind speed with regard to individual soil types are required.

In the region where the Earth's surface and subsurface are seasonally frozen, dust emissions begin to occur (early spring) when only the surface is melted and dry but not the subsurface just beneath a thin surface layer. The thermal inertia at the representative depth is still affected by the frozen soil, but the surface radiative temperature is highly positive in degrees Celsius due to the dried surface, which is suitable for dust emission (in other words, high erodibility) if the wind is necessarily strong. The temperature difference was up to at most 20°C (unpublished result). Dust emission is likely to occur in early spring when the meteorological conditions are likely to be windy; however, an empirical relationship between thermal inertia-derived soil moisture and threshold wind speed using observations during spring to early autumn [28] is difficult to apply because thermal inertia is likely to be underestimated in early spring possibly due to the large difference between the surface and the subsurface temperatures. There is no common formulation for thermal inertia-derived soil moisture with regard to the threshold wind speed in early spring or other seasons.

5.4. Water budget and management

Monitoring the spatial distribution of surface soil moisture over a wide agricultural area is required for optimal water management. An example presented by Minecapllia et al. [30] showed the spatial distribution of thermal inertia over a small-scale cultivated field using airborne thermal images taken in the daytime and nighttime.

Root zone soil moisture has been examined in several studies [50, 59], using a thermal inertia procedure with the FRM applied to the soil water transport and data assimilation procedures, respectively. All of the studies noted that the initial values of the root zone soil moisture were significant for reducing the simulation error. It was noted that the FRM applied to soil moisture was not straight-forward like the soil temperature because of the nonlinearity in soil water transport that representatively appeared in the Richards equation [34, 59]. Various processes of water transport in soil such as infiltration, redistribution, and vapor transport should be improved [60].

The precise spatial resolution of satellite LST is better used for coinciding topography or land use on approximately a 1-km scale. Overlaying or assimilating thermal inertia-derived soil moisture over a common scale of topography or land use in the range of a watershed should contribute to the water budget estimation (discharge, infiltration, and evapotranspiration) when precipitation is known. If agricultural land use is resolved at a 1-km resolution, it is practically suitable for estimating thermal inertia and its applications using M2018 with GCOM-C LST.

6. Conclusions

A vast number of studies have been proposed for estimating soil moisture, and thermal inertia-based methods have been improving for the last five decades. These methods are based on the thermal diffusion equation combined with sinusoidal function boundary conditions at the Earth's surface and are constant at an infinite depth. There are two solutions for retrieving thermal inertia of which both use satellite thermal-infrared-based LST. One uses the Fourier series expansion, and the other employs the force-restore method. There are advantages and disadvantages in their formulation, calculation procedures, time, and adjustment schemes; however, both solutions are essentially the same in principle. The parameterization for converting thermal inertia to soil moisture is also important. Two ways to perform this parameterization have been proposed. One uses the relationships of soil moisture to the volumetric heat capacity and thermal conductivity, and the other is analogous to the Johansen type thermal conductivity model. The individual studies proposed so far combined a method for retrieving thermal inertia and parameterization for converting it to soil moisture. The accuracy of estimating surface soil moisture for individual studies was not significantly different. The current and future applications of thermal inertia-derived soil moisture are discussed. The thermal inertia approach will be competitive with the assimilation method combining microwave-based soil moisture and satellite data from the other wavelength bands (visible, near-infrared, and thermal-infrared). There are advantages and disadvantages to both approaches in regard to spatial resolution, sky conditions, and the dry down process. Issues to be tackled remain for dust emission processes, especially in relation to soil moisture. Regional water budgeting and management can be applied to arid land water resource and agricultural practice considering the fusion of other satellite data such as GSMaP precipitation.

Acknowledgements

This chapter is based on an aggregation of contributions from personal collaborations and research groups. This chapter is also partly supported by JSPS KAKENHI Grant Number JP 26289159.

Appendix

Details of the weighting factor K_i formulation in Section 3.

Details of the formulation of K_i are given as

$$K_i = \frac{1}{3}\sum_{j=1}^{3}\left[1 + \left(\frac{\lambda_i}{\lambda_0} - 1\right)g_{ij}\right]^{-1}, \tag{20}$$

where λ_0 is the conductivity of the medium and g_{ij} is a shape factor that takes into account the shape and orientation of the granules. Since $\sum_{j=1}^{3} g_{ij} = 1$ and a spheroidal shape is assumed $(g_{i1} = g_{i2})$, the above equation reduces to:

$$K_i = \frac{2}{3\left[1 + \left(\frac{\lambda_i}{\lambda_0} - 1\right)g_i\right]} + \frac{1}{3\left[1 + \left(\frac{\lambda_i}{\lambda_0} - 1\right)(1 - 2g_i)\right]}. \tag{21}$$

Author details

Dai Matsushima

Address all correspondence to: matsushima.dai@it-chiba.ac.jp

Chiba Institute of Technology, Narashino, Chiba, Japan

References

[1] van Wijik WR, de Vries DA. Periodic temperature variations in a homogeneous soil. In: van Wijik WR, editor. Physics of Plant Environment. Amsterdam: North-Holland Publ. Co.; 1963. pp. 103-143

[2] Deardorff JW. Efficient prediction of ground surface temperature and moisture, with inclusion of a layer of vegetation. Journal of Geophysical Research. 1978;**83**:1889-1903. DOI: 10.1029/JC083iC04p01889

[3] Price JC. Thermal inertia mapping: A new view of the earth. Journal of Geophysical Research. 1977;**82**:2582-2590. DOI: 10.1029/JC082i018p02582

[4] Price JC. The potential of remotely sensed thermal infrared data to infer surface soil moisture and evaporation. Water Resources Research. 1980;**16**:787-795. DOI: 10.1029/WR016i004p00787

[5] Price JC. Estimating surface temperatures from satellite thermal infrared data—A simple formulation for the atmospheric effect. Remote Sensing of Environment. 1983;**13**:353-361. DOI: 10.1016/0034-4257(83)90036-6

[6] Price JC. On the analysis of thermal infrared imagery: The limited utility of apparent thermal inertia. Remote Sensing of Environment. 1985;**18**:59-73. DOI: 10.1016/0034-4257(85)90038-0

[7] Pratt DA, Ellyett CD. The thermal inertia approach to mapping of soil moisture and geology. Remote Sensing of Environment. 1979;**8**:151-168. DOI: 10.1016/0034-4257(79)90014-2

[8] van de Griend AA, Camillo PJ, Gurney RJ. Discrimination of soil physical parameters, thermal inertia, and soil moisture from diurnal surface temperature fluctuations. Water Resources Research. 1985;**21**:997-1009. DOI: 10.1029/WR021i007p00997

[9] Sobrino JA, El Kharraz MH. Combining afternoon and morning NOAA satellites for thermal inertia estimation: 1. Algorithm and its testing with hydrologic atmospheric pilot experiment-Sahel data. Journal of Geophysical Research. 1999;**104**:9445-9453. DOI: 10.1029/1998JD200109

[10] Maltese A, Bates PD, Capodici F, Cannarozzo M, Ciraolo G, La Loggia G. Critical analysis of thermal inertia approaches for surface soil water content retrieval. Hydrological Sciences Journal. 2013;**58**:1144-1161. DOI: 10.1080/02626667.2013.802322

[11] Xue Y, Cracknell AP. Thermal inertia mapping: From research to operation. In: Cracknelll AP, Vaughan RA, editors. Proceedings of the 18th Annual Conference of the Remote Sensing Society; 15–17 September 1992; University of Dundee. Nottingham, UK: Remote Sensing Society; 1992. pp. 471-480

[12] Xue Y, Cracknell AP. Advanced thermal inertia modelling and its application: Modelling emissivity of the ground. In: Proceedings of the 25th International Symposium on Remote Sensing and Global Environmental Change; 4–8 April 1993; Graz, Austria. Ann Arbor: ERIM; 1993. pp. II-121-II-122

[13] Xue Y, Cracknell AP. Advanced thermal inertia modelling. International Journal of Remote Sensing. 1995;**16**:431-446. DOI: 10.1080/01431169508954411

[14] Cracknell AP, Xue Y. Dynamic aspects of surface temperature from remotely-sensed data using advance thermal inertia model. International Journal of Remote Sensing. 1996;**17**: 2517-2532. DOI: 10.1080/01431169608949090

[15] Matsushima D. Estimating regional distribution of surface heat fluxes by combining satellite data and a heat budget model over the Kherlen River Basin, Mongolia. Journal of Hydrology. 2007;**333**:86-99. DOI: 10.1016/j.jhydrol.2006.07.028

[16] Matsushima D, Asanuma J, Kaihotsu I. Thermal inertia approach using a heat budget model to estimate the spatial distribution of surface soil moisture over a semi-arid grassland in Central Mongolia. Journal of Hydrometeorology. 2018;**19**:245-265. DOI: 10.1175/JHM-D-17-0040.1

[17] Dickinson RE. The force-restore model for surface temperatures and its generalizations. Journal of Climate. 1988;**1**:1086-1097. DOI: 10.1175/1520-0442(1988)001<1086:TFMFST>2.0.CO;2

[18] Hirota T, Pomeroy JW, Granger RJ, Maule CP. An extension of the force-restore method to estimating soil temperature at depth and evaluation for frozen soils under snow. Journal of Geophysical Research. 2002;**107**:ACL 11-1-ACL 11-10. DOI: 10.1029/2001JD001280

[19] Matsushima D, Kimura R, Shinoda M. Soil moisture estimation using thermal inertia: Potential and sensitivity to data conditions. Journal of Hydrometeorology. 2012;**13**:638-648. DOI: 10.1175/JHM-D-10-05024.1

[20] Matsushima D, Sensui Y, Ryuzaki T, Misaka I, Ando K, Yokoyama H, Narita K. A study on thermal environmental assessment in a street space: 1. Estimating heat storage in the subsurface and buildings using data from surface temperature measurements. In: Proceedings of the Annual Meeting of Architectural Institute of Japan; 26–29 August 2009; Sendai. Tokyo: Architectural Institute of Japan; 2009. pp. 729-730

[21] Tanaka S, Takeda H, Tsuchiya T, Iwata T, Terao M. Architectural Environmental Engineering. 3rd ed. Tokyo: Inoue Shoin; 2006. p. 324

[22] Anderson MC, Kustas WP, Norman JM, Hain CR, Mecikalski JR, Schultz L, et al. Mapping daily evapotranspiration at field to continental scales using geostationary and polar orbiting satellite imagery. Hydrology and Earth System Sciences. 2011;**15**:223-239. DOI: 10.5194/hess-15-223-2011

[23] Wu P, Shen H, Zhang L, Göttsche FM. Integrated fusion of multi-scale polar-orbiting and geostationary satellite observations for the mapping of high spatial and temporal resolution land surface temperature. Remote Sensing of Environment. 2015;**156**:169-181. DOI: 10.1016/j.rse.2014.09.013

[24] Scheidt S, Ramsey M, Lancaster N. Determining soil moisture and sediment availability at White Sands Dune Field, New Mexico, from apparent thermal inertia data. Journal of Geophysical Research. 2010;**115**:F02019. DOI: 10.1029/2009JF001378

[25] Schädlich S, Göttsche FM, Olesen FS. Influence of land surface parameters and atmosphere on METEOSAT brightness temperatures and generation of land surface temperature maps by temporally and spatially interpolating atmospheric correction. Remote Sensing of Environment. 2001;**75**:39-46. DOI: 10.1016/S0034-4257(00)00154-1

[26] Cai G, Xue Y, Hu Y, Wang Y, Guo J, Luo Y, Wu C, Zhong S, Qi S. Soil moisture retrieval from MODIS data in Northern China Plain using thermal inertia model. International Journal of Remote Sensing. 2007;**28**:3567-3581. DOI: 10.1080/01431160601034886

[27] Verhoef A. Remote estimation of thermal inertia and soil heat flux for bare soil. Agricultural and Forest Meteorology. 2004;**123**:221-236. DOI: 10.1016/j.agrformet.2003.11.005

[28] Matsushima D, Kimura R, Kurosaki Y, Shinoda M. A method for estimating the threshold wind speed as a function of soil moisture in a local scale using multiple models and data archives. (in Preparation)

[29] de Vries DA. Thermal properties of soils. In: Van Wijik WR, editor. Physics of Plant Environment. New York: Wiley; 1963. pp. 210-235

[30] Minacapilli M, Iovino M, Blanda F. High resolution remote estimation of soil surface water content by a thermal inertia approach. Journal of Hydrology. 2009;**379**:229-238. DOI: 10.1016/j.jhydrol.2009.09.055

[31] Lu S, Ren T, Gong Y, Horton R. An improved model for predicting soil thermal conductivity from water content at room temperature. Soil Science Society of America Journal. 2007;**71**:8-14. DOI: 10.2136/sssaj2006.0041

[32] Dong Y, McCartney JS, Lu N. Critical review of thermal conductivity models for unsaturated soils. Geotechnical and Geological Engineering. 2015;**33**:207-221. DOI: 10.1007/s10706-015-9843-2

[33] Ma AN, Xue Y. A study of remote sensing information model of soil moisture. In: Proceedings of the 11th Asian Conference on Remote Sensing, I. 15–21 November 1990; Beijing; 1990. pp. P-11-1-P-11-5

[34] Noilhan J, Planton S. A simple parameterization of land surface processes for meteorological models. Monthly Weather Review. 1989;**117**:536-549. DOI: 10.1175/1520-0493(1989)117<0536:ASPOLS>2.0.CO;2

[35] McCumber MC, Pielke RA. Simulation of the effects of surface fluxes of heat and moisture in a mesosclae numerical model: 1. Soil layer. Journal of Geophysical Research. 1981;**86**:9929-9938. DOI: 10.1029/JC086iC10p09929

[36] Clapp RB, Hornberger GM. Empirical equations for some soil hydraulic properties. Water Resources Research. 1978;**14**:601-604. DOI: 10.1029/WR014i004p00601

[37] Cosby BJ, Hornberger GM, Clapp RB, Ginn TR. A statistical exploration of the relationships of soil moisture characteristics to the physical properties of soils. Water Resources Research. 1984;**20**:682-690. DOI: 10.1029/WR020i006p00682

[38] Johansen O. Thermal conductivity of soils [PhD thesis]. University of Trondheim; 1975

[39] Murray T, Verhoef A. Moving tiwards a more mechanistic approach in the determination of soil heat flux from remote measurements: I. A universal approach to calculate thermal inertia. Agricultural and Forest Meteorology. 2007;**147**:80-87. DOI: 10.1016/j.agrformet.2007.07.004

[40] Lu S, Ju Z, Ren T, Horton R. A general approach to estimate soil water content from thermal inertia. Agricultural and Forest Meteorology. 2009;**149**:1693-1698. DOI: 10.1016/j.agrformet.2009.05.011

[41] Minacapilli M, Cammalleri C, Ciraolo G, D'Asaro F, Iovino M, Maltese A. Thermal inertia modelling for soil surface water content estimation: A laboratory experiment. Soil Science Society of America Journal. 2012;**76**:92-100. DOI: 10.2136/sssaj2011.0122

[42] Lu Y, Horton R, Zhang X, Ren T. Accounting for soil porosity improves a thermal inertia mode for estimating surface soil water content. Remote Sensing of Environment. 2018;**212**:79-89. DOI: 10.1016/j.rse.2018.04.045

[43] Holmes TRH, Hain CR, Anderson MC, Crow WT. Cloud tolerance of remote-sensing technologies to measure land surface temperature. Hydrology and Earth System Sciences. 2016;**20**:3263-3275. DOI: 10.5194/hess-20-3263-2016

[44] Kondo J, Watanabe T. Studies on the bulk transfer coefficients over a vegetated surface with a multilayer energy budget model. Journal of the Atmospheric Sciences. 1992;**49**: 2183-2199. DOI: 10.1175/1520-0469(1992)049<2183:SOTBTC>2.0.CO;2

[45] Norman JM, Kustas WP, Humes KS. A two-source approach for estimating soil and vegetation energy fluxes in observations of directional radiometric surface temperature. Agricultural and Forest Meteorology. 1995;**77**:263-293. DOI: 10.1016/0168-1923(95)02265-Y

[46] Nelder JA, Mead R. A simplex method for function minimization. Computer Journal. 1965;**7**:308-313. DOI: 10.1093/comjnl/7.4.308

[47] Sawada Y, Koike T, Walker JP. A land data assimilation system for simultaneous simulation of soil moisture and vegetation dynamics. Journal of Geophysical Research – Atmospheres. 2015;**120**:5910-5930. DOI: 10.1002/2014JD022895

[48] Sawada Y, Koike T. Towards ecohydrological drought monitoring and prediction using a land data assimilation system: A case study on the horn of Africa drought (2010–2011). Journal of Geophysical Research – Atmospheres. 2016;**121**:8229-8242. DOI: 10.1002/2015JD024705

[49] Sawada Y, Koike T, Aida K, Toride K, Walker JP. Fusing microwave and optical satellite observations to simultaneously retrieve surface soil moisture, vegetation water content, and surface soil roughness. IEEE Geoscience and Remote Sensing. 2017;**55**:6195-6206. DOI: 10.1109/TGRS.2017.2722468

[50] Bandara R, Walker JP, Rüdiger C, Merlin O. Towards soil property retrieval from space: An application with disaggregated satellite observations. Journal of Hydrology. 2015;**522**: 582-593. DOI: 10.1016/j.jhydrol.2015.01.018

[51] Shellito PJ, Small EE, Colliander A, Bindlish R, Cosh MH, Berg AA, Bosch DD, Caldwell TG, Goodrich DC, McNairn H, Prueger JH, Starks PJ, van der Velde R, Walker JP. SMAP soil moisture drying more rapid than observed in situ following rainfall events. Geophysical Research Letters. 2016;**43**:8068-8075. DOI: 10.1002/2016GL069946

[52] Kubota T, Shige S, Hashizume H, Aonashi K, Takahashi N, Seto S, Hirose M, Takayabu YN, Ushio T, Nakagawa K, Iwanami K, Kachi M, Okamoto K. Global precipitation map using satellite-borne microwave radiometers by the GSMaP project: Production and validation. IEEE Geoscience and Remote Sensing. 2007;**45**:2259-2275. DOI: 10.1109/TGRS.2007.895337

[53] Shao Y. Physics and Modelling of Wind Erosion. 2nd Revised and Expanded ed. Berlin: Springer; 2008. 452p

[54] Sharratt BS, Collins HP. Wind Erosion potential influenced by tillage in an irrigated potato–sweet corn rotation in the Columbia Basin. Agronomy Journal. 2018;**110**:842-849. DOI: 10.2134/agronj2017.12.0681

[55] Fécan F, Marticorena B, Bergametti G. Parametrization of the increase of the aeolian erosion threshold wind friction velocity due to soil moisture for arid and semi-arid areas. Annales Geophysicae. 1999;**17**:149-157. DOI: 10.1007/s00585-999-0149-7

[56] McKenna-Neuman C, Nickling WG. A theoretical and wind tunnel investigation of the effect of capillary water on the entrainment of sediment by wind. Canadian Journal of Soil Science. 1989;**69**:79-96. DOI: 10.4141/cjss89-008

[57] Sharratt BS, Vaddella VK, Feng G. Threshold friction velocity influenced by wetness of soils within the Columbia Plateau. Aeolian Research. 2013;**9**:175-182. DOI: 10.1016/j.aeolia.2013.01.002

[58] Sharratt BS, Vaddella V. Threshold friction velocity of crusted windblown soils in the Columbia Plateau. Aeolian Research. 2014;**15**:227-234. DOI: 10.1016/j.aeolia.2014.08.002

[59] Verstraeten WW, Veroustraete F, van der Sande CJ, Grootaers I, Feyen J. Soil moisture retrieval using thermal inertia, determined with visible and thermal spaceborne data, validated for European forests. Remote Sensing of Environment. 2006;**101**:299-314. DOI: 10.1016/j.rse.2005.12.016

[60] Or D, Lehmann P, Shahraeeni E, Shokri N. Advances in soil evaporation physics—A review. Vadose Zone Journal. 2013;**12**. DOI: 10.2136/vzj2012.0163

Chapter 3

Soil Moisture Retrieval from Microwave Remote Sensing Observations

Hongquan Wang

Additional information is available at the end of the chapter

http://dx.doi.org/10.5772/intechopen.81476

Abstract

This chapter mainly describes the vegetated soil moisture retrieval approaches based on microwave remote sensing data. It will be comprised of three topics: (1) SAR polarimetric decomposition is to model the full coherency matrix as a summation of the surface, dihedral, and volume scattering mechanisms. After removing the volume scattering component, the soil moisture is estimated from the surface and dihedral scattering components. Particularly, various dynamic volume scattering models will be critically reviewed, allowing the readers to select the appropriate one to capture the complex variations of the volume scattering mechanism with crop phenological growth. (2) Radiative transfer model is to express the radar backscattering coefficient as the incoherent summation of different scattering components. Hereby, we will review the water cloud model and its several extensions for enhanced soil moisture retrieval. (3) Compared to the active radar, the passive radiometer possesses high temporal resolution but coarse spatial resolution. The third topic is dedicated to review the microwave emission models and the active-passive combined approaches, in the context of Soil Moisture and Ocean Salinity (SMOS) and Soil Moisture Active and Passive (SMAP) missions.

Keywords: soil moisture, polarimetric decomposition, radiative transfer model, microwave emission model

1. Introduction

Soil moisture is an important factor influencing the food supply to human beings at the small scale, and also an essential climate change variable that needs to be monitored at a large scale. In order to estimate the spatiotemporal dynamics of the soil moisture, the Soil Moisture and

Ocean Salinity (SMOS) satellite was launched in 2009, followed by the Soil Moisture Active and Passive (SMAP) satellite launched in 2015 although the radar component failed to send the signal back. These two missions used the microwave band, considering the dependence of the emissivity on the target dielectric constant and the penetration ability at long frequency. The microwave is found to be an appropriate frequency for monitoring the soil moisture, as it is not influenced by the cloud, and can operate day/night. Nevertheless, the passive radiometer signal is limited by the coarse spatial resolution. In contrast, the radar signal is characterized by higher spatial resolution and longer revisit time. Thus, it is appropriate to employ the radar signals for the soil moisture at a scale of agricultural fields. The polarimetric radars such as the ALOS PALSAR and RADARSAT-2 provide a full coherency or covariance matrix, which contain more information than the single-channel radar system. The PolSAR allows to extract the scattering mechanisms, which are useful for the land classification and geophysical parameter retrievals.

The soil moisture retrieval from the microwave remote sensing data is mainly influenced by the vegetation, surface roughness, and soil texture. However, over the agricultural fields, the crop characteristics vary with the phenological growth, leading to the complexity to model the vegetation influences on the soil moisture retrieval. For instance, the quality of the polarimetric soil moisture retrieval approach is highly dependent on the volume scattering model, which is used to remove the vegetation scattering contribution in the full polarized radar signal. To address this issue, several adaptive volume scattering models were developed at L-band [1] and C-band [2] for tracking the dynamic of crop growth. Both the retrieval accuracy and retrieval rate are enhanced by the dynamic volume scattering models. In contrast, in the radiative transfer models, the vegetation effect is often simulated by the vegetation optical depth, which is subsequently related to the vegetation water content and the normalized differential vegetation index (NDVI).

Within this context, this chapter provides a review of the model-based polarimetric decomposition approach, radiative transfer models, and combined active-passive methods for soil moisture retrieval over the vegetated agricultural fields. Particularly, different adaptive volume scattering models for the polarimetric decomposition are compared, and the optimal application conditions are drawn for the soil moisture retrieval. This chapter gives readers an overview of the soil moisture retrieval models at microwave band.

2. Soil and vegetation parameters influencing the microwave signals

SAR system transmits polarimetric waves toward the targets and receives the backscattering signals after the interaction with ground and ground targets. This technique is of great importance for agricultural managers to monitor the soil properties and surface conditions of the agricultural fields. For example, the retrievals of soil status information from SAR can be used to identify areas at risk of erosion by water and wind. Thus, in this study, we propose to investigate soil moisture and surface roughness as two important parameters describing the properties of bare agricultural fields. First of all, we propose to describe the parameterization of soil moisture and surface roughness.

2.1. Soil moisture

Soil is considered as three-phase materials: liquid phase, solid particles, and air phase. The liquid phase can be categorized into two types: the bound water and free water. Bound water is comprised of the water molecules contained in the first few molecular layers surrounding the soil particles. They are tightly held by the soil particles due to the influence of osmotic and matric forces [3, 4]. As the distance away from the soil particle surface increases, the matric forces decrease; thus the water molecules located far from the soil particle are able to move within the soil medium, which is referred as free water. Nevertheless, the criterion to separate bound water and free water is to some extent arbitrary. The amount of bound water located in the first few layers is determined by the surface area of the soil particles, which depends on the distribution of soil particle size. According to the distribution of soil particle size, different soils can be categorized into different soil textures. The solid particles are the second phase, which make up the soil skeleton. The void space between soil particles may be full of water if the soil is saturated or may be full of air if the soil is dry or may be partially saturated. The water percent hold in the soil particles is considered as soil moisture. There exist several expressions for soil moisture representation, and the frequently used approaches are the volumetric soil moisture mv and gravimetric soil moisture mg. The relationship between the volumetric soil moisture mv and gravimetric soil moisture mg is established by the water density ρ_w and total mass density ρ_b: $mv = mg \cdot \rho_b/\rho_w,$ where mv is measured using time-domain reflectometry (TDR) and mg is used to calibrate the TDR measurements. Nevertheless, the soil texture must be taken into account in order to determine the soil capability for stocking water.

2.1.1. Soil texture

Soil texture is reported to have great effects on the dielectric behaviors over the entire microwave frequency range and is most significant at frequencies around 5 GHz [5]. Different soil textures can be qualitatively classified used both in field and laboratory measurements based on their physical properties. The classes are distinguished by the "textural feel" which can be further clarified by separating the relative proportions of sand, silt, and clay using grading sieves. The classes are then used to determine the crop suitability and to approximate the soil responses to environmental conditions [6]. Different soil elements which determine the specific soil texture are separated and based on the specific ranges of particle diameter d [7]:

- The smallest particles are clay particles with $d < 0.002$ mm.
- The next smallest particles are silt particles with $0.002 \text{ mm} < d < 0.05$ mm.
- The largest particles are sand particles with $d > 0.05$ mm.

Soil texture classification is based on relative combination of sand, silt, and clay. Clay particles are microscopic in size and are highly plastic at moist condition. The presence of silt and/or clay creates a fine texture soil, which impedes water and air movements. Sand-sized particles are visible with the naked eye.

2.1.2. *Soil permittivity*

The complex dielectric constant describes the behaviors of nonconductor in the electrical field. A number of factors affect the dielectric constant, such as wave frequency, temperature, and salinity of the matter. The dielectric constant represents the maximum capability to store, absorb, and conduct electric energy for a given matter. It is a measure of the medium response to the electromagnetic wave and is defined as $\varepsilon_a = \varepsilon_a^{'} - i\varepsilon_a^{''} = \varepsilon_0\left(\varepsilon_r^{'} - i\varepsilon_r^{''}\right)$, where ε_a represents the absolute complex permittivity, $\varepsilon_a^{'}$ and $\varepsilon_a^{''}$ are the real and imaginary parts of ε_a, and $\varepsilon_0 = 8.85 \cdot 10^{-12}(\mathrm{F/m})$ is the vacuum permittivity. $\varepsilon_r^{'}$ is referred as the relative permittivity and considered as the dielectric constant of the specific medium. $\varepsilon_r^{''}$ is referred as the absorption capabilities of the medium and is relative to its conductivity and dielectric loss. For most natural medium, the condition $\varepsilon_r^{'} \gg \varepsilon_r^{''}$ is satisfied.

The relative dielectric constant of water is around 80, much larger than those of solid soil (2–5) and air (around 1) [3]. Hence, the permittivity of natural soils which are mixtures of three matters is influenced largely by water content. It is viable to measure the dielectric constant in order to infer the soil water content. However, under very dry soil conditions, the real part of the dielectric constant $\varepsilon_r^{'}$ ranges from 2 to 4, and the imaginary part $\varepsilon_r^{''}$ is below 0.05 [8]. This low dielectric constant results in the soil moisture underestimation by TDR instruments, because the water is tightly bounded to the surface of soil particle, and it causes only a relatively small increase of soil permittivity which cannot be detected by TDR. On the contrary, as the water content continues to increase, above the specific transition soil moisture value (free water becomes dominant in soils), the soil permittivity will increase rapidly.

In addition, assuming the propagating wave attenuates exponentially in soils, the penetrating depths δ_p of microwave into the soil (skin depth) can be calculated as [9, 10]

$$\delta_p = \frac{\lambda\sqrt{\varepsilon_r^{'}}}{2\pi\varepsilon_r^{''}} \tag{1}$$

It is noted that as the wavelength increases, the penetrating depth increases, as shown in **Figure 1** for L-, C-, and X-band, respectively. Meanwhile, for a given wavelength, the penetrating depth decreases as soil moisture increases.

2.1.3. *Conversion between soil moisture and soil dielectric constant*

Topp model: The soil permittivity is expressed as a three-order polynomial function in Topp model [4], which is only available for wave frequency between 20 MHz and 1 GHz:

$$\varepsilon_r^{'} = 3.03 + 9.3mv + 146mv^2 - 76.7mv^3 \tag{2}$$

http://dx.doi.org/10.5772/intechopen.81476

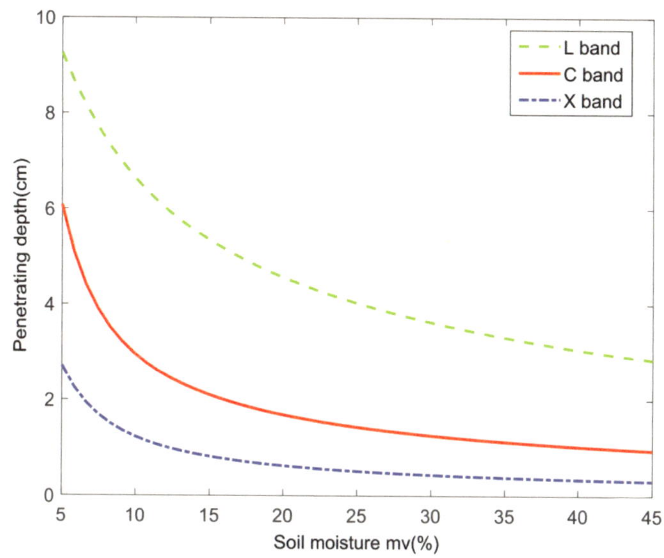

Figure 1. The penetrating depth in terms of radar frequency and soil moisture.

Inversely, the soil moisture is deduced from the soil permittivity measurements by a similar three-order polynomial equation:

$$mv = -5.3 \times 10^{-2} + 2.92 \times 10^{-2}\varepsilon_r' - 5.5 \times 10^{-4}\varepsilon_r'^2 + 4.3 \times 10^{-6}\varepsilon_r'^3 \quad (3)$$

This model does not consider the imaginary part of dielectric constant, and the main restriction is that the used frequency must be less than 1 GHz. The in situ soil moisture measurements using TDR are based on this model.

Hallikainen model: A more applicative conversion model is proposed by [5], and the soil permittivity is modeled as a function of soil moisture and soil texture in a two-order polynomial form:

$$\varepsilon_r = (a_0 + a_1S + a_2C) + (b_0 + b_1S + b_2C)mv + (c_0 + c_1S + c_2C)mv^2 \quad (4)$$

where a_i, b_i, and c_i (i = 1, 2, 3) are the complex coefficients for difference wave frequency between 1.4 and 18 GHz. Thus, both the real and imaginary parts of soil permittivity can be modeled. The S and C represent the percentage of silt and clay components, respectively.

Mironov model: The soil dielectric constant depends on the soil water content, temperature, texture, and wavelength. In the past decades, the semiempirical models in [4, 11] were mainly used for both the active and passive microwave remote sensing of soil moisture. Furthermore, Mironov dielectric model [12] considers the difference between the bound water and free water in the soil layers, which is found to be better for soil moisture retrieval at L-band.

2.2. Surface roughness

Besides the soil moisture, the surface roughness is another important factor that affects the backscattering SAR signature, because it determines how the incidence wave interacts with the surface. There exist several ways to describe the natural surface roughness, and two frequently used methods are mentioned here: the fractal geometry theory and the statistical description.

2.2.1. Fractal description

The fractal geometry theory was introduced in [13] to describe the complicated surface roughness structure, especially for the irregular and fragmented soil structures. This surface roughness description approach is proved to be suitable for natural soil because of its self-similarity, no matter what the surface scale is. In addition, many basic natural physical processes generate fractal surface; thus fractal structure is quite common in natural environment.

The fractal models describe the local structure of the soil surface using one parameter, the fractal dimension *D*, ranged from 1 to 2. The higher the fractal dimension, the rougher the surface. One of the frequently used fractal approaches is the Brownian model [14, 15] for a limited fractal profile. In this model, the surface profile height h(x) at location x is considered to be a fractional Brownian function: For any x and $\Delta x > 0$, the increase of surface height $h(x + \Delta x) - h(x)$ follows a Gaussian distribution with mean value zero and variance $A\Delta x^{2H}$. Consequently, the expected value of the surface elevation increase is derived as

$$E[\mathrm{h(x + \Delta x) - h(x)}] = 2\int_0^\infty \frac{u}{\sqrt{2\pi A}\Delta x^H} \exp\left(\frac{-u^2}{2A\Delta x^{2H}}\right)\mathrm{du} = \Delta x^H E[\mathrm{h(x+1) - h(x)}] \tag{5}$$

where A is the variance of the normal distribution $h(x + 1) - h(x)$ and H is the Hurst exponent constant ranged from 0 to 1. The equation is rewritten in logarithm format in order to resolve H:

$$\log[\mathrm{h(x + \Delta x) - h(x)}] = \mathrm{H}\log(\Delta \mathrm{x}) + \log[\mathrm{h(x+1) - h(x)}] \tag{6}$$

In this equation, the parameter H equals to the slope of $\log[\mathrm{h(x + \Delta x) - h(x)}]$ in terms of $\log(\Delta x)$. It is calculated by using minimum RMSE method [16]. Consequently, the fractal dimension D can be obtained directly from H by the relationship $D = 2 - H$.

2.2.2. Statistical description

The second approach to describe the surface roughness is from the statistical point of view. There are two parameters to describe the statistical variations of the surface height relative to a reference surface: the standard deviation of the surface height s is to quantify the vertical roughness, while the correlation length l (with autocorrelation function) is to characterize the horizontal roughness [9, 17].

Suppose a surface in the x-y plane and the height of point (x, y) are assumed to be z(x, y) above the x-y plane. A representative surface with dimensions Lx and Ly is segmented statistically, which is centered at the original point.

The average height of the surface is given by

$$\overline{z} = \frac{1}{L_x L_y} \int_{-L_x/2}^{L_x/2} \int_{-L_{y/2}}^{L_{y/2}} z(x,y)dxdy \tag{7}$$

and the second moment is given by

$$\overline{z}^2 = \frac{1}{L_x L_y} \int_{-L_x/2}^{L_x/2} \int_{-L_{y/2}}^{L_{y/2}} z^2(x,y)dxdy \tag{8}$$

Consequently, the standard deviation of the surface height within the area Lx X Ly is defined as

$$s = \sqrt{(z^2) - \overline{z}^2} \tag{9}$$

The formulation above can be reduced to a discrete condition. The surface profiles are digitized into discrete values $z_i(x_i)$ at spacing rate Δx which is satisfied the criterion $\Delta x < 0.1\lambda$ as described in [3]. The standard deviation s for discrete condition is formulated as

$$s = \sqrt{\frac{\sum_{i=1}^{N} (z_i)^2 - N(\hat{z})^2}{N-1}} \tag{10}$$

where $\hat{z} = \frac{\sum_{i=1}^{N} z_i}{N}$ is the mean surface height and N is the number of samples.

For the horizontal surface roughness description, the surface autocorrelation function (ACF) has to be determined. The autocorrelation function ρ characterizes the independence of two points at a distance $\sqrt{\xi^2 + \zeta^2}$:

$$\rho(\xi, \zeta) = \frac{\int_{-Lx/2}^{Lx/2} \int_{-Ly/2}^{Ly/2} z(x,y)z(x+\xi, y+\zeta)dxdy}{\int_{-Lx/2}^{Lx/2} \int_{-Ly/2}^{Ly/2} z^2(x,y)dxdy} \tag{11}$$

In the discrete case, the autocorrelation function for a spatial displacement $x_i = (j - 1)\Delta x$ is defined as

$$\rho(\xi) = \frac{\sum_{i=1}^{N+1-j} z_j z_{j+i-1}}{\sum_{i=1}^{N} {z_i}^2} \quad (12)$$

where z_{j+i-1} is a point with the spatial displacement x_i from the point z_i. The surface correlation length l is then defined as the displacement x_i, under which the $\rho(x_i)$ between the two points equals 1/e. The correlation length characterizes the statistical independence of two points. In case that the distance between two points is larger than l, their heights can be considered statistically independent from each other. For very smooth surface, the correlation length is toward infinity.

2.2.3. *Wave interaction with the surface roughness*

Furthermore, the effective surface roughness observed by SAR system depends on microwave wavelength. For instance, a given surface that appears smooth in L-band may seem rough in C-band. The relative surface roughness status (compared with wavelength) affects the surface scattering behaviors:

- For the smooth surface, the angular radiation pattern of the reflected wave is modeled as a delta function which is centered about the specular direction.
- For the medium roughness surface, the angular radiation pattern is comprised of coherent component and incoherent component. The coherent component is radiated in the specular direction even though its magnitude is smaller than over the smooth surface. The incoherent scattering component consists of energy scattered in all directions, but its magnitude is smaller than that of the coherent component.
- For the rough surface, the radiation pattern seems like a Lambertian surface, comprised of only incoherent scattering.

Thus, in the electromagnetic models, the effective vertical and horizontal surface roughness is given in terms of the production with EM wave number ($k = 2\pi f/c$): ks and kl. It is obvious that ks and kl are decreasing with increasing wavelength. Consequently, as shown in **Figure 2**, the surface roughness is one of the important factors that determine the electromagnetic wave response from bare soil.

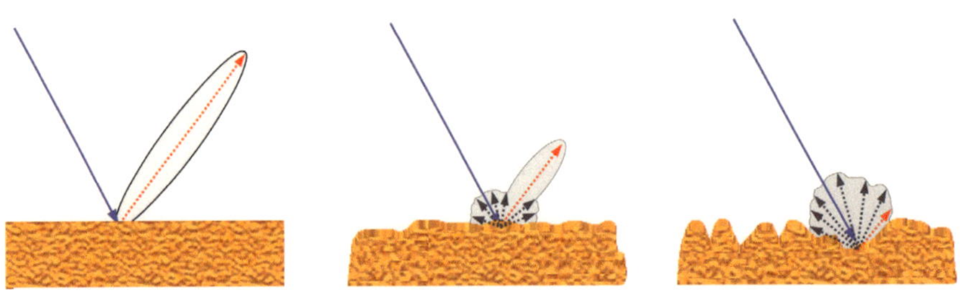

Figure 2. Scattering patterns determined by surface roughness.

2.2.4. *Bragg phenomenon*

Except the surface roughness and soil moisture, the row direction also influences the backscattering SAR wave from the bare agricultural soils, because it induces the Bragg phenomenon. Bragg resonance is a type of coherent scattering, which is present in some agricultural fields due to the plowing or other row structures' tillage. The resonance occurs in case that the distance between radar and each of the periodic structures has an additional phase difference of $\lambda/2$ in the slant-range direction. Under this condition, the additional phase shift is 2π, and the signals will add in phase.

2.3. Vegetation

Vegetation has two effects on the radar signal: (1) attenuate the backscattering from the underlying soils and (2) produce the volume scattering adding to the radar signal. These two effects increase the complexity of soil moisture retrieval from microwave signal. The vegetation attenuation and scattering effects were parameterized by the vegetation scattering albedo and optical depth, which are related to the vegetation water content or leaf area index.

A. Vegetation optical depth τ is linked to the vegetation water content through b parameter:

$$\tau = b\mathrm{VWC} \tag{13}$$

The b parameter depends on the crop type, structure, and growth stage and microwave polarization. The vegetation water content is often obtained from the NDVI. Alternatively, τ can be obtained from the LAI through a linear relationship:

$$\tau = b_1\mathrm{LAI} + b_2 \tag{14}$$

The b_1 and b_2 are assumed to be dependent on the vegetation type.

B. Vegetation scattering albedo ω is set to be zero or a low value in the passive radiometer analysis.

3. Polarimetric decomposition for soil moisture estimation

Depending whether the sensor generates the microwave by itself, the microwave remote sensing can be categorized into the active and passive, which are reviewed separately. Polarimetric SAR is a coherent active microwave remote sensing system, providing backscattering signals in quad-polarization states with fine spatial resolution. Unlike the optical remote sensing, the SAR system monitors the earth using a side-look geometry, resulting in the issues of overlap, shadow, and forth short. Furthermore, at the microwave bands, the signals are sensitive to the permittivity and the structure of the targets. Thus, the interpretation and modeling of the SAR data differ from those of optical domain. The SAR system generates the microwave, so that it operates regardless the light and day/night and clear/cloudy conditions.

This is particularly interesting for monitoring the soil moisture over the area frequently covered by the cloud.

3.1. Decomposition theories

3.1.1. Polarimetric SAR data expression

The microwave scattering process over the ground can be formulated $E_S = [S]E_I$, where the Sinclair matrix [S] relates the incident wave E_I to the scattering wave E_S. Thus, the polarimetric SAR data extracted as [S] includes the target dielectric and structural properties:

$$[S] = \begin{bmatrix} S_{HH} & S_{HV} \\ S_{VH} & S_{VV} \end{bmatrix} \tag{15}$$

where S_{HH} and S_{VV} are the co-polarized scatterings, and S_{HV} and S_{VH} represent the cross-polarized scattering power. They are all complex numbers. For the monostatic case of back-scattering, the satisfied reciprocity results in $S_{HV} = S_{VH}$. This format of [S] matrix is considered as single-look data suffering from the speckle effect, as no averaging process is performed.

However, the natural targets dynamically vary with time, requiring a statistical description such as the second-order moment approach. In order to extract more polarimetric information such as the correlation between different polarimetric channels, the Pauli and Lexicographic vectors are constructed from the [S] matrix, respectively:

$$k = [S_{HH} + S_{VV},\ S_{HH} - S_{VV},\ 2S_{HV}]^T \tag{16}$$

$$\Omega = \left[S_{HH},\ \sqrt{2}S_{HV},\ S_{VV}\ \right]^T \tag{17}$$

From the Pauli and Lexicographic vectors, the coherency matrix [T] and the covariance matrix [C] are obtained by $[T] = \langle k \cdot k^{*T} \rangle$ and $[C] = \langle \Omega \cdot \Omega^{*T} \rangle$, where the symbol $\langle\rangle$ means the temporal or spatial averaging to reduce the randomness of the polarimetric images. In the monostatic condition (transmitter and receiver in the same location), they are expressed as

$$[T3] = \begin{bmatrix} T_{11} & T_{12} & T_{13} \\ T_{12}^* & T_{22} & T_{23} \\ T_{13}^* & T_{23}^* & T_{33} \end{bmatrix} = 0.5 \begin{bmatrix} \left\langle |S_{HH} + S_{VV}|^2 \right\rangle & \langle (S_{HH} + S_{VV})(S_{HH} - S_{VV})^* \rangle & 2\langle (S_{HH} + S_{VV})S_{HV}^* \rangle \\ \langle (S_{HH} + S_{VV})^*(S_{HH} - S_{VV}) \rangle & \left\langle |S_{HH} - S_{VV}|^2 \right\rangle & 2\langle (S_{HH} - S_{VV})S_{HV}^* \rangle \\ 2\langle (S_{HH} + S_{VV})^* S_{HV} \rangle & 2\langle (S_{HH} - S_{VV})^* S_{HV} \rangle & 4\left\langle |S_{HV}|^2 \right\rangle \end{bmatrix} \tag{18}$$

$$[C3] = \begin{bmatrix} C_{11} & C_{12} & C_{13} \\ C_{12}^* & C_{22} & C_{23} \\ C_{13}^* & C_{23}^* & C_{33} \end{bmatrix} = \begin{bmatrix} \langle |S_{HH}|^2 \rangle & \sqrt{2}\langle S_{HH}S_{HV}^* \rangle & \langle S_{HH}S_{VV}^* \rangle \\ \sqrt{2}\langle S_{HH}^*S_{HV} \rangle & 2\langle |S_{HV}|^2 \rangle & \sqrt{2}\langle S_{HV}S_{VV}^* \rangle \\ \langle S_{HH}^*S_{VV} \rangle & \sqrt{2}\langle S_{HV}^*S_{VV} \rangle & \langle |S_{VV}|^2 \rangle \end{bmatrix}$$

The polarimetric decompositions are often done on the coherency matrix [T3] and the covariance matrix [C3], which can be converted between each other via unitary transformation. However, the elements of the [T3] matrix are physically convenient. For instance, the T_{11}, T_{22}, and T_{33} can be used to approximate the surface, dihedral, and volume scattering powers.

3.1.2. Eigen-based decomposition

Both [T3] and [C3] matrices are characterized by nonnegative eigenvalues and orthogonal eigenvector. The classical decomposition approach proposed by Cloude and Pottier relies on the eigenanalysis on the [T3] matrix. The scattering mechanism and the corresponding relative power were quantified by the eigenvector (T_i) and eigenvalues (λ_i), respectively:

$$[T3] = \lambda_1 T_1 + \lambda_2 T_2 + \lambda_3 T_3 \tag{19}$$

From the eigenvalues and eigenvectors, the entropy H and α angle are defined to characterize the randomness of the scattering scene and the dominant scattering mechanism:

$$H = \sum_{i=1}^{3} -p_i \log_3 p_i, \quad \alpha = \sum_{i=1}^{3} p_i \operatorname{acos}(|e_{i1}|) \quad \text{and} \quad p_i = \lambda_i / \sum_{i=1}^{3} \lambda_i \tag{20}$$

In addition, the scattering anisotropy A is introduced to discriminate the ambiguous case of $H > 0.7$:

$$A = (\lambda_2 - \lambda_3)/(\lambda_2 + \lambda_3) \tag{21}$$

These polarimetric parameters are used to describe the scattering mechanisms under a variety of scenarios. However, in Baghdadi et al. [18], the sensitivity of entropy and α angle to soil moisture and surface roughness is analyzed, indicating insignificant response of these polarimetric parameters to the soil characteristics at C-band.

3.1.3. Model-based decomposition

Under the assumption of reflection symmetry (zero correlation between the co- and cross-polarization channels), the Freeman-Durden decomposition models the covariance matrix [C3] as the incoherent summation of the surface, dihedral, and volume scattering components.

In order to be consistent with previous eigen-based approach, we express the Freeman-Durden decomposition based on [T3] matrix [19]:

$$[T3] = \begin{bmatrix} T_{11} & T_{12} & 0 \\ T_{12}^{*} & T_{22} & 0 \\ 0 & 0 & T_{33} \end{bmatrix} = f_s \begin{bmatrix} 1 & \beta^{*} & 0 \\ \beta & |\beta|^2 & 0 \\ 0 & 0 & 0 \end{bmatrix} + f_d \begin{bmatrix} |\alpha|^2 & \alpha & 0 \\ \alpha^{*} & 1 & 0 \\ 0 & 0 & 0 \end{bmatrix} + f_v \begin{bmatrix} V_{11} & 0 & 0 \\ 0 & V_{22} & 0 \\ 0 & 0 & V_{33} \end{bmatrix} \quad (22)$$

The surface component is modeled using the simple Bragg model. The polarimetric parameter $\beta = \frac{R_H - R_V}{R_H + R_V}$ and $f_s = 0.5|R_H + R_V|^2$ are constructed from the Bragg scattering coefficients:

$$R_H = \frac{\cos\theta - \sqrt{\varepsilon - \sin^2\theta}}{\cos\theta + \sqrt{\varepsilon - \sin^2\theta}}, R_V = \frac{(\varepsilon - 1)\left(\sin^2\theta - \varepsilon\left(1 + \sin^2\theta\right)\right)}{\left(\varepsilon\cos\theta + \sqrt{\varepsilon - \sin^2\theta}\right)^2} \quad (23)$$

The dihedral component is developed from the Fresnel coefficients of the orthogonal dielectric planes between the plant stalks and the underlying soils. The scattering amplitude $f_d = 0.5|R_{SH}R_{TH} + R_{SV}R_{TV}e^{i\psi}|^2$ and polarization ratio $a = \frac{R_{SH}R_{TH} - R_{SV}R_{TV}e^{i\psi}}{R_{SH}R_{TH} + R_{SV}R_{TV}e^{i\psi}}$ are related to the Fresnel coefficients of soil and plant:

$$R_{jH} = \frac{\cos\theta_j - \sqrt{\varepsilon_i - \sin^2\theta_j}}{\cos\theta_j + \sqrt{\varepsilon_i - \sin^2\theta_j}} \text{ and } R_{jV} = \frac{\varepsilon_i\cos\theta_j - \sqrt{\varepsilon_i - \sin^2\theta_j}}{\varepsilon_i\cos\theta_j + \sqrt{\varepsilon_i - \sin^2\theta_j}} \quad (24)$$

where j represents soil (S) or plant (T). In the dihedral geometric configuration, the incidence angle over soil θ_S and over the plant θ_T is supplementary ($\theta_S + \theta_T = \frac{\pi}{2}$).

The vegetation volume is simulated by the dipole with a uniform statistical distribution. Consequently, the volume component is derived as

$$[V] = fv \begin{bmatrix} 0.5 & 0 & 0 \\ 0 & 0.25 & 0 \\ 0 & 0 & 0.25 \end{bmatrix} \quad (25)$$

The Freeman-Durden model is firstly fitted to the forest scenario, and it is reported to be effective to discriminate the forest and deforest areas.

3.2. Soil moisture retrieval using polarimetric decomposition techniques

3.2.1. Model-based decomposition

The polarimetric soil moisture retrieval can be conducted based on the model-based decomposition, in which the soil dielectric constant is related to the surface scattering component through the Bragg scattering model and to the dihedral component through the combined Fresnel scattering model. Nevertheless, in the past decades, the model-based polarimetric decompositions were mainly applied to the image classification, target detection by analyzing the scattering mechanisms. Hajnsek et al. [19] proposed to estimate the soil moisture from the L-band polarimetric decomposition. In their approach, after removing the volume component from the full signature, the soil moisture is retrieved from the surface and dihedral scattering component, respectively.

For the surface scattering component, the polarimetric parameter β is related to the soil moisture and incidence angle (**Figure 3**). Unlike the traditional radar backscattering coefficients which are more sensitive to soil moisture at low incidence angle condition, the polarimetric parameter β is more sensitive to the soil moisture at high incidence angle. Thus, depending on the incidence angle ranges of the radar data, the traditional direct backscattering approach or the advanced polarimetric approach is preferable. In Hajnsek et al. [19], the surface scattering component is adapted by replacing the Bragg model with the X-Bragg model in order to take the surface roughness effect into account.

In contrary to the surface scattering component, the dihedral scattering component is influenced by both the soil and vegetation dielectric constants. Thus, two equations were required to decouple the soil and vegetation contributions on the dihedral component, in order to extract the soil moisture from it. In the literature [19, 20], the parameter α and f_d are used to construct an equation system, from which the soil and vegetation dielectric constants are solved. Nevertheless, the vegetation dielectric constant is not furthermore considered, as the main purpose of this chapter is to estimate the soil moisture from microwave remote sensing data.

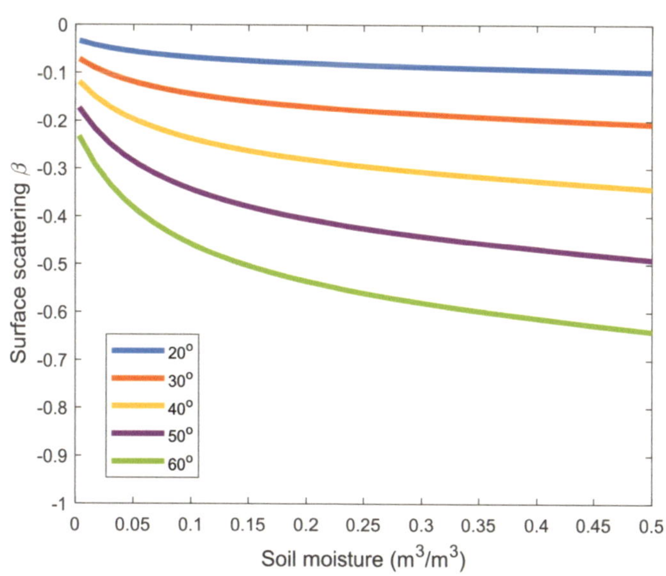

Figure 3. Sensitivity of surface scattering parameter β to soil moisture.

Figures 4, 5 plot the α and f_d in terms of soil and vegetation dielectric constants:

- Parameter α is more sensitive to soil moisture when the incidence angle is less than 45°; otherwise, it is more sensitive to vegetation dielectric constants.
- For the f_d parameter, the sensitivity to soil moisture is the same between the low and high incidence angles, while the absolute value of f_d is different.

The dihedral scattering component is complementary to the surface component, increasing the overall retrieval rate. The surface scattering component which is the function of only soil dielectric constant is generally easier for the soil moisture retrieval than the dihedral component which is the function of both soil and vegetation dielectric constants. However, for some crop types such as canola and wheat, the significant dihedral scattering power at the early phenological stages contributes largely to the soil moisture [21]. There is a limitation in the dihedral component at incidence angle around 45°, when the soil and vegetation dielectric constants are not possible to be decoupled from each other.

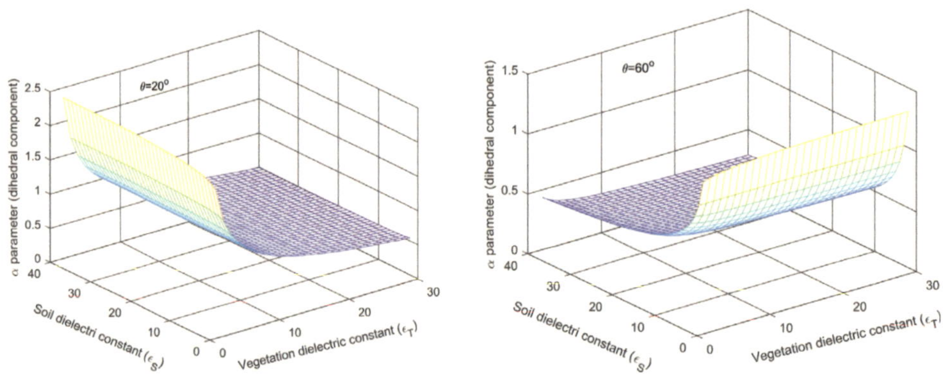

Figure 4. Sensitivity of dihedral parameter alpha to soil and vegetation dielectric constants under low and high incidence angles.

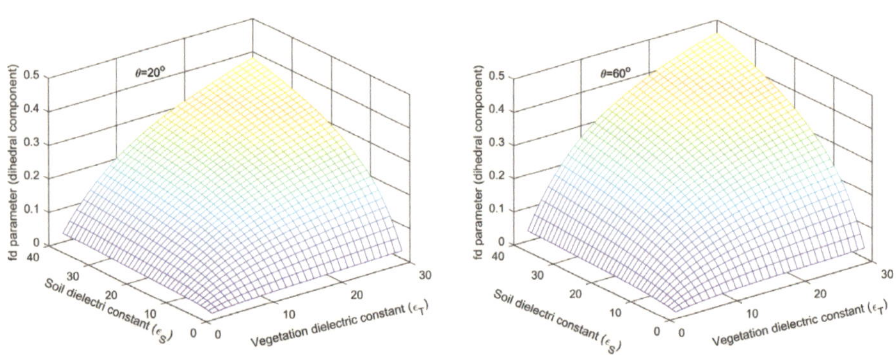

Figure 5. Sensitivity of alpha parameter f_d to soil and vegetation dielectric constants under low and high incidence angles.

It is in the consensus that the most challenging issue is the modeling of the volume scattering component. With the crop growth, the shape and crop structures vary dynamically, which makes the unique volume coherency matrix fail to capture the high complexity of the crop growth. In order to analyze this issue, Hajnsek et al. [19] compared several volume scattering formulations. One is the flexible volume model in Yamaguchi et al. [22], where the crops are described in vertical, random, and horizontal orientations. The volume coherency matrix was derived considering the dipoles with different orientation angle distribution widths. The parameter $Pr = 10\log10(VV)/10\log10(HH)$ is used to determine the dominant orientation:

$$T_V = \frac{fv}{30}\begin{bmatrix} 15 & 5 & 0 \\ 5 & 7 & 0 \\ 0 & 0 & 8 \end{bmatrix}, T_R = fv\begin{bmatrix} 0.5 & 0 & 0 \\ 0 & 0.25 & 0 \\ 0 & 0 & 0.25 \end{bmatrix}, T_H = \frac{fv}{30}\begin{bmatrix} 15 & -5 & 0 \\ -5 & 7 & 0 \\ 0 & 0 & 8 \end{bmatrix} \quad (26)$$

Another volume coherency matrix is proposed by narrowing the dipole orientation angle around radar line of sight. However, for all the volume models in [19], the corresponding soil moisture retrieval results indicate an underestimation for the wheat and corn fields, while an over-/underestimation for the rape fields. So far, there is no universal volume coherency matrix which performs well for all the crop types and the whole phenological development stages.

Furthermore, Jagdhuber et al. [20] developed an L-band polarimetric decomposition for the multiangular soil moisture retrieval over the agricultural fields covered by low vegetation. In the study, the multiangular observation was conducted by three flight lines over the same area. The effects of microwave extinction and phase shift on the surface and dihedral scattering component were accounted. For each pixel, multiple β, α, and fd were obtained for a joint retrieval process. The soil moisture retrieval obtained an RMSE ranging from 0.06 to 0.08 m^3/m^3.

Recently, the hybrid decomposition which combines the model-based and eigen-based decompositions is used for the soil moisture retrieval [1]. After extracting the volume scattering component using the model-based approach, the remaining ground scattering component is decomposed again using the eigen-based approach in order to better discriminate the surface and dihedral scattering mechanisms, taking advantages of the orthogonality of the eigenvector. This avoids the assumption of the dominant scattering mechanism in the ground component, in the original Freeman-Durden decomposition approach [23].

In addition, the deorientation process is accounted before conducting the polarimetric decomposition, to reduce the fluctuation due to the random orientation angle of each pixel. This was done by minimizing the cross-polarization power [24]. After the deorientation process, the pixel with different orientation angles will result in the same decomposition results. Wang et al. [25] studied effectivity of the deorientation on the polarimetric soil moisture, indicating that the surface scattering component is significantly enhanced, as a result of the deorientation process. The increase in the surface scattering power is assumed to benefit the soil moisture retrieval. This is understandable, as the surface component is a function of the soil characteristics, while the dihedral component is complicated due to the coupling between the soil and vegetation dielectric constants. Three different polarimetric decompositions (Freeman-Durden, Hajnsek, and An) were compared for the soil moisture retrieval. However, the performances depend on the crop types and phenological stages, and none of them can perform well for all the crop types and the whole growth stages. The Hajnsek decomposition is better for the early growth stage, while the An decomposition is overperformed for the crop's later development season. The Freeman decomposition obtained better results on the corn fields with sparse planting density. Furthermore, the incidence angle normalization is conducted on the polarimetric parameters (β, α, and fd) to reduce the incidence angle effect on the soil moisture retrieval.

Similar to the idea of X-Bragg model which rotates the Bragg surface around radar line of sight, the extended Fresnel model was developed for the dihedral scattering component [26]. It is achieved by rotating the soil plane of the dihedral component around the radar line of sight, to introduce the surface roughness effect on the dihedral component. Unlike the introduction of the surface roughness in the dihedral component in Hajnsek et al. [19], which did not change the matrix rank, the dihedral coherency matrix obtained in the extended Fresnel model increases the matrix rank from 1 to 3. Thus, both the amplitude and phase of the dihedral component have been changed.

3.2.2. Eigen-based decomposition

The eigenvalues and eigenvectors of [T] matrix were computed to construct the polarimetric parameters for characterization of the scattering mechanisms. However, the currently eigen-based decomposition is mainly limited for soil moisture retrieval over the bare soils. The first one is the X-Bragg model [27], introducing the surface roughness effect into the Bragg model by rotating the soil plane around the radar light of sight. In order to estimate the soil moisture, the X-Bragg model relates the entropy H and α angle to the soil dielectric constant. A lookup table is established to determine the soil dielectric constant from the data-derived entropy and α angle. In addition, the surface roughness is derived from the polarimetric anisotropy parameter.

Furthermore, under the assumption of the reflection symmetry, the polarimetric parameters which are dominated by only the soil moisture or the surface roughness were constructed from the eigenvalue and eigenvector of the coherency matrix. According to Allain [28], the analytical eigenvalues is derived as

$$\begin{aligned}
\lambda_{1nos} &= 0.5\left(< |S_{HH}|^2 > + < |S_{VV}|^2 > + \sqrt{\left(< |S_{HH}|^2 > - < |S_{VV}|^2 > \right)^2 + 4 < |S_{HH}S_{VV}^*|^2 >} \right) \\
\lambda_{2nos} &= 0.5\left(< |S_{HH}|^2 > + < |S_{VV}|^2 > - \sqrt{\left(< |S_{HH}|^2 > - < |S_{VV}|^2 > \right)^2 + 4 < |S_{HH}S_{VV}^*|^2 >} \right) \\
\lambda_{3nos} &= 2 < |S_{HV}|^2 >
\end{aligned} \tag{27}$$

where the sign *nos* denotes no order in size. The corresponding analytical eigenvectors can be derived as

$$e_1 = \frac{1}{\sqrt{2\left[\left(|S_{HH}|^2 - |S_{VV}|^2 + \sqrt{\Delta}\right)^2 + 4\left|S_{HH}S_{VV}^*\right|^2\right]}} \begin{bmatrix} 2S_{HH}S_{VV}^* + \left(|S_{HH}|^2 - |S_{VV}|^2 + \sqrt{\Delta}\right) \\ 2S_{HH}S_{VV}^* - \left(|S_{HH}|^2 - |S_{VV}|^2 + \sqrt{\Delta}\right) \\ 0 \end{bmatrix}$$

$$e_2 = \frac{1}{\sqrt{2\left[\left(|S_{HH}|^2 - |S_{VV}|^2 - \sqrt{\Delta}\right)^2 + 4\left|S_{HH}S_{VV}^*\right|^2\right]}} \begin{bmatrix} 2S_{HH}S_{VV}^* + \left(|S_{HH}|^2 - |S_{VV}|^2 - \sqrt{\Delta}\right) \\ 2S_{HH}S_{VV}^* - \left(|S_{HH}|^2 - |S_{VV}|^2 - \sqrt{\Delta}\right) \\ 0 \end{bmatrix}$$

$$e_3 = \begin{bmatrix} 0 \\ 0 \\ 1 \end{bmatrix} \tag{28}$$

with $\Delta = \left(|S_{HH}|^2 - |S_{VV}|^2\right)^2 + 4\left|S_{HH}S_{VV}^*\right|^2$. Based on the eigenvalues in reflection symmetry conditions, three polarimetries SERD, DERD, and SDERD are defined to characterize the difference among three scattering mechanisms (single bounce, double bounce, and multiple scattering):

$$\begin{aligned} \mathrm{SERD} &= \frac{\lambda_s - \lambda_m}{\lambda_s + \lambda_m} \\ \mathrm{DERD} &= \frac{\lambda_d - \lambda_m}{\lambda_d + \lambda_m} \\ \mathrm{SDERD} &= \frac{\lambda_s - \lambda_d}{\lambda_s + \lambda_d} \end{aligned} \tag{29}$$

where $\lambda_s = \lambda_{1nos}$ and $\lambda_s = \lambda_{2nos}$ if $a_1 < a_2$. In contrary, $\lambda_s = \lambda_{2nos}$ and $\lambda_s = \lambda_{1nos}$ if $a_1 > a_2$. The $\lambda_m = \lambda_{3nos}$ holds on in all cases. It is reported [10] that SERD is suitable to characterize vegetation, while DERD is appropriate to quantify the surface roughness. SDERD can be applied to discriminate between bare and sight vegetation soils.

In order to find a polarimetric parameter which is sensitivity to soil moisture, the α_1 from the first eigenvector is derived as

$$\alpha_1 = \arctan\left(\frac{2\sigma_{HHVV} - \left(\sigma_{VVVV} - \sigma_{HHHH} + \sqrt{(\sigma_{VVVV} - \sigma_{HHHH})^2 + 4|\sigma_{HHVV}|^2}\right)}{2\sigma_{HHVV} + \left(\sigma_{VVVV} - \sigma_{HHHH} + \sqrt{(\sigma_{VVVV} - \sigma_{HHHH})^2 + 4|\sigma_{HHVV}|^2}\right)}\right) \tag{30}$$

In Allain [28], the IEM model is used to simulate the backscattering coefficients. It is found that α_1 tends to be invariable with respect to the radar frequency higher than 8 GHz. At such high frequency, the α_1 is approximated using the IEM model as

$$\lim_{\substack{f \to hight \\ frequency}} \alpha_1 = \arctan\left(\frac{2f_{hh}f_{vv}^* - \left(|f_{vv}|^2 - |f_{hh}|^2 + \sqrt{\left(|f_{vv}|^2 - |f_{hh}|^2\right)^2 + 4|f_{hh}f_{vv}^*|^2}\right)}{2f_{hh}f_{vv}^* + \left(|f_{vv}|^2 - |f_{hh}|^2 + \sqrt{\left(|f_{vv}|^2 - |f_{hh}|^2\right)^2 + 4|f_{hh}f_{vv}^*|^2}\right)}\right) \tag{31}$$

where the f_{hh} and f_{hh} are the parameters in the IEM model. In this case, the α_1 is independent of surface roughness and mainly depends on the soil dielectric constant. The potential of α_1 for the soil moisture retrieval is investigated in Baghdadi et al. [18] using the C-band RADARSAT-2 data, indicating it is possible to discriminate two soil moisture levels or provide necessary a priori information to enhance the accuracy of soil moisture retrieval.

3.2.3. *Hybrid decomposition*

The eigen-based decomposition is more empirically used for soil moisture retrieval, as it is inherently a mathematical approach. In contrast, the model-based decomposition based on the Bragg and Fresnel scattering models is more physically used. Recently, the combination between the model-based and eigen-based decompositions results in the hyper-decomposition [1]. Firstly, the volume scattering component is removed using the model-based decomposition. Then, the remaining ground scattering is decomposed using the eigen-based decomposition. This process overcomes the requirement of assumption on the dominant surface or dihedral scattering mechanism in the ground component (in that case, we need to assume the β or α to be constant in order to solve the undetermined equation system).

Furthermore, as the vegetation shape and structure vary with the phenological growth, the limited volume scattering model is not sufficient to capture this complex variability. Thus, the dynamic volume scattering is developed [1], which is suitable for the entire crop phenological cycle:

$$[\mathrm{Tv}] = \frac{f_v}{2 + 2A_p^2}\begin{bmatrix} V_{11} & V_{12} & 0 \\ V_{12}^* & V_{22} & 0 \\ 0 & 0 & V_{33} \end{bmatrix} \tag{32}$$

$$V_{11} = \left(\mathrm{A}_p + 1\right)^2$$

$$V_{12} = \left(A_p^2 - 1\right)^2 \mathrm{sinc}(2\Delta\varphi)$$

$$V_{22} = 0.5\left(A_p - 1\right)^2 (1 + \mathrm{sinc}(4\Delta\varphi))$$

$$V_{33} = 0.5\left(A_p - 1\right)^2 (1 - \mathrm{sinc}(4\Delta\varphi))$$

The parameters A_p and $\Delta\phi$ were used to characterize the vegetation shape and its distribution width, respectively. With the dynamic volume model and hyper-decomposition approach, the soil moisture estimation obtained an inversion rate higher than 95% and RMSE from 4.0 to 4.4 m^3/m^3. In addition, for the covariance matrix, a generalized volume scattering model is proposed in [29] to quantify the vegetation scattering using the cosine-square distribution. Although the formulation varies from one to another study, the main idea relies on the characterization of the vegetation shape and orientation using the minimum number of parameters.

However, the model-based polarimetric decomposition for the soil moisture retrieval is mainly valid at L-band. When it comes to C-band, the surface roughness condition is beyond the valid range of Bragg ($ks < 0.3$) or X-Bragg model ($ks < 1$). In order to overcome this limitation, Huang et al. [2] first proposed a C-band polarimetric decomposition for the slight vegetation condition. In their approach, the surface scattering component is simulated using the IEM model, while the volume scattering component is formulated using the first-order sine and cosine functions for the vertical and horizontal orientations. Finally, a RMSE of 6.12 m^3/m^3 is obtained for the soil moisture retrieval using the C-band RADARSAT-2 dataset.

4. Radiative scattering model

The soil moisture retrieval is performed using either physical or empirical models. We introduced below the application of integral equation model (IEM) and Oh model over the bare soil and the water cloud model (WCM) over the vegetated condition.

4.1. IEM model

The IEM model can be used to simulate the backscattering coefficients from incidence angle θ and soil parameters (surface roughness ks, correlation length kl, and soil moisture mv). Two surface roughness conditions (Gaussian or exponential) are considered to compute the corresponding backscattering coefficients. Regarding the applicability of IEM model, some studies show reasonable agreements between measurements and the model [30, 31]. However, the disagreements between measurements and model predictions are frequently observed [32–36], because the IEM model backscattering behavior depends on the autocorrelation function (ACF). Furthermore, the measurement of correlation length l is difficult to be accurate enough, since this parameter is dependent on the profiler length as well as the number of repetition in the surface roughness measurements [37, 38].

To overcome the limitations of IEM model, Baghdadi proposed in [35, 36, 39] a calibration procedure for HH, VV, and HV polarization channels, respectively. It is assumed that the disagreements between IEM model and actual datasets are due to the selection of autocorrelation function and the in situ correlation length measurements. Therefore, after fitting a large set of experiment datasets, a calibration parameter l_c is built for different polarization channels at different incidence angles in order to take the place of measured correlation length l. The calibration parameter l_c given in [40] for C-band is described with respect to RMS surface height s and incidence angle θ:

- For HH polarization: $l_c = 0.162 + 3.006(\sin 1.23\theta)^{-1.194}s$
- For VV polarization: $l_c = 1.281 + 0.134(\sin 0.19\theta)^{-1.59}s$
- For HV polarization: $l_c = 0.9157 + 1.2289(\sin 0.1543\theta)^{-0.3139}s$

By replacing the measured correlation length with this calibration parameter, the agreement between the IEM model simulation and actual radar measurement is reported to be improved [35, 40].

4.2. Oh model

The Oh model is established based on theoretical scattering models [9], scatterometer measurements, and airborne polarimetric SAR datasets (in L-, C-, and X-band, respectively) under different roughness and soil moisture conditions at incidence angles ranging from 10 to 70°. This model relates the co-polarized ratio $p = \sigma^0_{HH}/\sigma^0_{VV}$ and the cross-polarized ratio $q = \sigma^0_{HV}/\sigma^0_{VV}$ and absolute σ^0_{HV} to soil parameters (including s, l, ε) and radar system parameters (including the wave number k and local incidence angle θ).

4.3. Water cloud model

As a first-order radiative transfer solution, the WCM model expresses the total backscattering signals as the summation of surface and volume scattering components, $\sigma^0_{total} = \Gamma^2\sigma^0_{surface} + \sigma^0_{volume}$. The surface scattering can be modeled using the bare soil moisture model such as the previous IEM and Oh models. The vegetation two-way attenuation on the surface scattering power is modeled by $\Gamma^2 = \exp(-2\tau/\cos\theta)$.

The vegetation layer is assumed to be comprised of homogenous water particles with a uniform distribution, and volume scattering component can be expressed from vegetation scattering albedo and optical depth such as $\sigma^0_{volume} = 0.75\omega(1-\Gamma^2)\cos\theta$. Accounting the polarization leads to the following empirical volume power [8]:

$$\begin{aligned}\sigma_{volume}^{VV-pola} &= \sigma_{volume}^{HH-pola} = 0.74\omega\left[1 + 0.54\omega\tau - 0.24(\omega\tau)^2\right]\left[1 - \exp\left(-2.12\tau/\cos\theta\right)\right]\cos\theta \\ \sigma_{volume}^{HV-pola} &= \omega\left[0.044\omega\tau - 0.018(\omega\tau)^2 + 0.006(\omega\tau)^3\right]\left[1 - \exp\left(-11.7\tau/\cos\theta\right)\right]\cos\theta\end{aligned} \tag{33}$$

At the moderate or high frequency such as C- and X-bands, the dihedral scattering is negligible. However, at low frequency such as L-band, the dihedral scattering component must be accounted, which can be quantified as [8]

$$\begin{aligned}\sigma_{dihedral}^{HH-pola} &= 1.9\omega\left[1 + 0.9\omega\tau + 0.4(\omega\tau)^2\right]\left[1 - \exp\left(-1.93\tau/\cos\theta\right)\right]\exp\left(-1.37\tau^{1.12}/\cos\theta\right) \\ &\quad \exp\left(-0.84(\mathrm{ks})^2\cos\theta\right)|\mathrm{R}_{HH}|^2\cos\theta \\ \sigma_{dihedral}^{HV-pola} &= 0.013\omega\left[1 + 7.85\omega\tau + 7.9(\omega\tau)^2\right]\left[1 - \exp\left(-9.62\tau/\cos\theta\right)\right]\exp\left(-1.02\tau^{1.38}/\cos\theta\right) \\ &\quad \exp\left(-2.9(\mathrm{ks})^2\cos\theta\right)\left(|\mathrm{R}_{HH}|^2 + |\mathrm{R}_{VV}|^2\right)0.5\cos\theta \\ \sigma_{dihedral}^{VV-pola} &= 0\end{aligned} \tag{34}$$

5. Soil and vegetation emission model

To collect sufficient emitted energy at microwave bands, satellite radiometer uses large footprint, resulting in coarse spatial resolution. Based on the measured brightness temperature, two typical models are applied for the soil moisture retrieval: L-band Microwave Emission of the Biosphere (L-MEB) and Land Parameter Retrieval Model (LPRM). The former was mainly developed for the L-band such as the SMOS mission, while the latter was mostly used at high frequency but can be also applied to L-band. All these models were based on a simple τ-ω model for vegetation covered lands.

Vegetation effect: The τ-ω model is formulated to account for the vegetation effect on the brightness temperature. It simulates the TB at polarization p (h or v) as the incoherent summation of (*i*) the soil emission attenuated by the vegetation, (*ii*) vegetation direct upwelling microwave emission, and (*iii*) vegetation downwelling emissions which are reflected by the soils and attenuated by the vegetation itself:

$$TB_p = E_p\gamma_p T_{soil} + (1-\omega)\left(1-\gamma_p\right)\mathrm{T}_{vege} + (1-\omega)\left(1-\gamma_p\right)\mathrm{T}_{vege}\mathrm{R}_p\gamma_p \tag{35}$$

where T_{soil} and T_{vege} are soil and vegetation effective temperatures, respectively. The soil emissivity $Ep = (1 - Rp)$ is computed from the soil reflectivity (Rp). The vegetation attenuation on the soil emission is modeled through a vegetation transmissivity γ_p which is a function of the optical depth τ_p and incidence angle θ:

$$\gamma_p = e^{-\tau_p/\cos(\theta)} \quad \text{and} \quad \tau_p = b_p \cdot VWC \tag{36}$$

At L-band, the vegetation scattering albedo ω is assumed to be close to zero and independent of the polarization and incidence angle [41].

Surface roughness effect: assuming the surface scattering over the interface between soil and air, the rough soil reflectivity R_p was obtained from the smooth surface reflectivity r_p:

$$R_p = \left[(1 - Q) \cdot r_p + Q \cdot r_q\right] \cdot \exp\left(-H_r \cos^{N_p}(\theta)\right) \tag{37}$$

where r_p is the Fresnel coefficients for h and v polarizations:

$$\begin{aligned} r_h &= \left| \frac{\cos\theta - \sqrt{\varepsilon_s - \sin^2\theta}}{\cos\theta + \sqrt{\varepsilon_s - \sin^2\theta}} \right|^2 \\ r_v &= \left| \frac{\varepsilon_s \cos\theta - \sqrt{\varepsilon_s - \sin^2\theta}}{\varepsilon_s \cos\theta + \sqrt{\varepsilon_s - \sin^2\theta}} \right|^2 \end{aligned} \tag{38}$$

The parameter Q quantifies the polarization mixing degree due to the surface roughness and is neglected at L-band [42, 43]. Np represents the dependence of roughness on incidence angle. Furthermore, the effective surface roughness parameter H_r is associated to the measured surface roughness in a conventional way:

$$H_r = 4k^2 s^2 \tag{39}$$

with wave number k and surface RMS height s. However, the clear general relationship between H_r and measured surface roughness is still uncertain. In the literature, different empirical relationships were established to link the H_r parameter to surface RMS height and the autocorrelation length [44]. The H_r parameter is also found to be influenced by soil moisture, but it is reported to be mainly valid for the sandy soils [42].

6. Joint active-passive microwave for soil moisture estimation

The radar signal comprised of the amplitude and phase is coherent and more influenced by the surface roughness and vegetation. In contrast, the radiometer signal is incoherent, reducing the influences from the surface roughness and vegetation. In addition, the radar signal is acquired with high spatial resolution at the cost of narrow swath range, while the radiometer signal has a frequent revisit cycle but coarse spatial resolution. In order to combine the advantages of the radar and radiometer signals, recent studies go into the soil moisture retrieval by a joint active-passive approach. In this context, the original objective of the SMAP mission is to monitor the soil moisture by the active-passive combination, although the radar component failed.

For the airborne platform, the active and passive signals can be obtained with a similar spatial resolution. The optimization process is conducted to match the microwave signals to the model output. For instance, the following cost function was constructed [45] by using both the radar and radiometer signals:

$$C(X) = 0.5\left[\sum\left|\frac{\sigma_{pq}^{data} - \sigma_{pq}^{simu}(X)}{\sigma_{pq}^{data}}\right|^2 + \gamma\cdot\sum\left|\frac{TB_p^{data} - TB_p^{simu}(X)}{TB_p^{data}}\right|^2\right] \quad (40)$$

where σ_{pq}^{data} and TB_p^{data} are the real data from the radar and radiometer, respectively. $\sigma_{pq}^{simu}(X)$ and $TB_p^{simu}(X)$ are the simulated radar and radiometer signals. The γ is a tuning parameter to balance the radar and radiometer signals in the optimization process. The increase of γ represents the enhanced contribution of the radiometer signals for the soil moisture retrieval. The airborne Passive-Active L-band Sensor (PALS) data were collected during the SMAPVEX12 and SMAPVEX16 campaigns, providing an opportunity to develop the active-passive soil moisture retrieval approaches.

For the spaceborne platform, such as the condition of the original SMAP mission, the radar and radiometer signals have different spatial resolutions. In this case, the radar signal with fine spatial resolution is used to disaggregate the radiometer signal with coarse resolution to obtain TB data with moderate resolution, considering the correlation between the radar and radiometer signals. Then, the emission model was applied to the disaggregated brightness temperature to retrieve the soil moisture at a moderate spatial resolution.

Author details

Hongquan Wang[1,2*]

*Address all correspondence to: hongquan.wang@usherbrooke.ca

1 College of Environmental and Resource Sciences, Zhejiang University, Hangzhou, China

2 CARTEL, University of Sherbrooke, Sherbrooke, QC, Canada

References

[1] Jagdhuber T, Hajnsek I, Papathanassiou KP. An iterative generalized hybrid decomposition for soil moisture retrieval under vegetation cover using fully polarimetric SAR. IEEE Journal of Selected Topics in Applied Earth Observations and Remote Sensing. 2015;**8**(8): 3911-3922

[2] Huang X, Wang J, Shang J. An integrated surface parameter inversion scheme over agricultural fields at early growing stages by means of C-Band polarimetric RADARSAT-2 imagery. IEEE Transactions on Geoscience and Remote Sensing. 2016;**54**(5):2510-2528

[3] Baver LD, Gardner WH, Gardner WR. Soil Physics. New York: Wiley; 1977

[4] Topp GC. Electromagnetic determination of soil water content: Measurements in coaxial transmission lines. Water Resource Research. 1980;**16**(3):574-582

[5] Hallikainen MT, Ulaby FT, Dobson MC, et al. Microwave dielectric behavior of wet soil-part 1: Empirical models and experimental observations. In: IEEE Transactions on Geoscience and Remote Sensing. Vol. 23; 1985. pp. 25-34

[6] Toogood JA. A simplified textural classification diagram. Canadian Journal of Soil Science. 1958;**38**:54-55

[7] Rowell DL. Soil Science: Methods and Applications. Harlow, UK: Longman Scientific & Technical; 1994

[8] Ulaby FT, Moore RK, Fung AK. Microwave Remote Sensing: Active and Passive, Vol. III – Volume Scattering and Emission Theory, Advanced Systems and Applications. Norwood, USA: Artech House; 1986

[9] Ulaby FT, Moore RK, Fung AK. Microwave Remote Sensing: Active and Passive, Vol. II – Radar Remote Sensing and Surface Scattering and Emission Theory. Norwood, US: Addison-Wesley Publishing Company, Advanced Book Program/World Science Division; 1982

[10] Daniel S. Analysis d'images SAR polarimetriques aeroportees pour l'estimation de parameters bio-physiques des sols agricoles. Universite de Rennes 1; 2009

[11] Dobson MC, Ulaby FT, Hallikainen MT, et al. Microwave dielectric behavior of wet soil-part II: Dielectric mixing models. IEEE Transactions on Geoscience and Remote Sensing. 1985;**GE-23**:35-46

[12] Mironov VL, Kosolapova LG, Fomin SV. Physically and mineralogically based spectroscopic dielectric model for moist soils. IEEE Transactions on Geoscience and Remote Sensing. 2009;**47**(7):2059-2070

[13] Mandelbrot BB. The Fractal Geometry of Nature. San Francisco, CA: Freeman; 1982

[14] Falconer K. Fractal Geometry, Mathematical Foundations and Applications. New York: John Wiley and Sons; 1990

[15] Zribi M, Ciarletti V, Taconet O, et al. Characterisation of the soil structure and microwave backscattering based on numerical three-dimensional surface representation: Analysis with a fractional Brownian model. Remote Sensing of Environment. 2000;**72**:159-169

[16] Jaynes ET. Probability Theory: The Logic of Science. Cambridge,UK: Cambridge University Press; 2003

[17] Hajnsek I. Inversion of Surface Parameters using Polarimetric SAR. Jena, Germany: Friedrich-Schiller-Universitat Jena; 2001

[18] Baghdadi N, Cresson R, Pottier E, et al. A potential use for the C-band polarimetric SAR parameters to characterize the soil surface over bare agriculture fields. IEEE Transaction on Geoscience and Remote Sensing. 2012;**50**(10):3844-3858

[19] Hajnsek I, Jagdhuber T, Schon H, et al. Potential of estimating soil moisture under vegetation cover by means of PolSAR. IEEE Transaction on Geoscience and Remote Sensing. 2009;**47**(2):442-454

[20] Jagdhuber T, Hajnsek I, Bronstert A, et al. Soil moisture estimation under low vegetation cover using a multi-angular polarimetric decomposition. IEEE Transaction on Geoscience and Remote Sensing. 2013;**51**(4):2201-2215

[21] Wang H, Magagi R, Goita K, et al. Evaluation of simplified polarimetric decomposition for soil moisture retrieval over vegetated agricultural fields. Remote Sensing. 2016; **8**(2):142

[22] Yamaguchi Y, Moriyama T, Ishido M, et al. Four-component scattering model for polarimetric SAR image decomposition. IEEE Transaction on Geoscience and Remote Sensing. 2005;**43**(8):1699-1706

[23] Freeman A, Durden SL. A three-component scattering model for polarimetric SAR data. IEEE Transaction on Geoscience and Remote Sensing. 1998;**36**(3):963-973

[24] An W, Cui Y, Yang J. Three-component model-based decomposition for polarimetric SAR data. IEEE Transaction of Geoscience and Remote Sensing. 2010;**48**(6):2732-2739

[25] Wang H, Magagi R, Goita K. Comparison of different polarimetric decompositions for soil moisture retrieval over vegetation covered agricultural area. Remote Sensing of Environment. 2017;**199**(1):120-136

[26] Jagdhuber T. An approach to extended Fresnel scattering for modeling of depolarizing soil-trunk double-bounce scattering. Remote Sensing. 2016;**8**(10). Article No. 818

[27] Hajnsek I, Pottier E, Cloude SR. Inversion of surface parameters from polarimetric SAR. IEEE Transactions on Geoscience and Remote Sensing. 2003;**41**(4):727-744

[28] Allain S. Caracterisation d'un sol nu a partir de donnees SAR polarimetriques etude multifrequentielle et multi-resolutions. Universite de Rennes 1; 2003

[29] Arii M, vanZyl JJ, Kim Y. Adaptive model-based decomposition of polarimetric SAR covariance matrices. IEEE Transactions on Geoscience and Remote Sensing. 2011;**49**(3): 1104-1113

[30] Shi JC, Wang J, Hsu AY, et al. Estimation of bare surface soil moisture and surface roughness parameter using L-band SAR image data. IEEE Transaction on Geoscience and Remote Sensing. 1997;**35**(5):1254-1266

[31] Bindlish R, Barros AP. Multifrequency soil moisture inversion from SAR measurements with the use of IEM. Remote Sensing of Environment. 2000;**71**:67-88

[32] Mattia F, Le Toan T, Souyrb J, et al. The effect of surface roughness on multifrequency polarimetric SAR data. IEEE Transactions on Geoscience and Remote Sensing. 1997;**35**: 954-966

[33] Zribi M, Taconet O, Hégarat-Mascle SL, et al. Backscattering behavior and simulation comparison over bare soils using SIR-C/X-SAR and ERASME 1994 data over Orgeval. Remote Sensing of Environment. 1997;**59**:256-266

[34] Baghdadi N, Gherboudj I, Zribi M, et al. Semi-empirical calibration of the IEM backscattering model using radar images and moisture and roughness field measurements. International Journal of Remote Sensing. 2004;**25**:3593-3623

[35] Baghdadi N, Holah N, Zribi M. Calibration of the integral equation model for SAR data in C-band and HH and VV polarizations. International Journal of Remote Sensing. 2006;**27**:4

[36] Baghdadi N, Chaaya JA, Zribi M. Semi-empirical calibration of the integral equation model for SAR Data in C-band and cross polarization using radar images and field measurements. IEEE Geoscience and Remote Sensing Letters. 2011;**8**(1):14-18

[37] Altese E, Bolognani O, Mancini M, et al. Retrieving soil moisture over bare soil from ERS-1 synthetic aperture radar data: Sensitivity analysis based on a theoretical surface scattering model and field data. Water Resources Research. 1996;**32**:653-661

[38] Oh Y, Kay YC. Condition for precise measurement of soil surface roughness. IEEE Transactions on Geoscience and Remote Sensing. 1998;**36**(2):691-695

[39] Baghdadi N, King C, Bonnifait L. An empirical calibration of the Integral Equation Model based on SAR data and soil parameters measurements. IEEE International Geoscience and Remote Sensing Symposium. Toronto, Canada. June 24-28 2002

[40] Dong L, Baghdadi N, Lu R. Validation of the AIEM through correlation length parameterization at field scale using radar imagery in a semi-arid environment. IEEE Geoscience and Remote Sensing Letters. 2013;**10**:461-465

[41] Wigneron JP, Parde M, Waldteufel P, et al. Characterizing the dependence of vegetation model parameters on crop structure, incidence angle, and polarization at L-band. IEEE Transactions on Geoscience and Remote Sensing. Feb. 2004;**42**(2):416-425

[42] Martens B, Lievens H, Colliander A, et al. Estimating effective roughness parameters of the L-MEB Model for soil moisture retrieval using passive microwave observations from SMAPVEX12. IEEE Transactions on Geoscience and Remote Sensing. 2015;**53**(7):4091-4103

[43] Wang JR, Neill PEO, Jackson TJ, et al. Multifrequency measurements of the effects of soil moisture, soil texture, and surface roughness. IEEE Transactions on Geoscience and Remote Sensing. 1983;**GE-21**(1):44-51

[44] Lawrence H, Wigneron JP, Demontoux F, et al. Evaluating the semiempirical H–Q model used to calculate the L-Band emissivity of a rough bare soil. IEEE Transactions on Geoscience and Remote Sensing. 2013;**51**(7):4075-4084

[45] Akbar R, Moghaddam M. A combined active-passive soil moisture estimation algorithm with adaptive regularization in support of SMAP. IEEE Transactions on Geoscience and Remote Sensing. 2015;**53**(6):3312-3324

Chapter 4

Correlation between TDR and FDR Soil Moisture Measurements at Different Scales to Establish Water Availability at the South of the Yucatan Peninsula

Judith Guadalupe Ramos Hernández,
Jesus Gracia-Sánchez,
Tania Patricia Rodríguez-Martínez and
José Adalberto Zuñiga-Morales

Additional information is available at the end of the chapter

http://dx.doi.org/10.5772/intechopen.81477

Abstract

The advantages that offer new techniques such as remote sensing to estimate soil moisture require local accurate measurements of this variable since these values are key to validate the estimated ones. The chapter analyses the performance to measure soil moisture using different sensors that correspond to different scales at the field. Sensors used were based on reflectometry, time and frequency, which were calibrated with gravimetric measurements. Additionally to have accurate soil moisture values, the idea is to have an operational system in a very complex ecosystem in order to see its influence to maintain the aguadas (small natural lagoons) at the south of the Yucatan Peninsula. These aguadas represent an important source of water in the region because the area presents shortage associated not only with the climate variation but also with high influence due to the type of soils (karst). Results demonstrated that the sensors tested were accurate particularly in the rainy season with some differences in the dry period, and also, the sensitivity of each device was determinant. Results will cover different areas from point to small regions (<4 km), since soil moisture data obtained could be extrapolated to different scales based on the climate, vegetation and type of soil, to compute the real water availability for the communities in the zone.

Keywords: soil moisture, DTR, FDR, aguadas, water availability

1. Introduction

Today, technological advances have favored a better understanding of the circumstances around natural phenomena, being necessary to explain, in order to comprehend the dynamic nature of the climate, all the interrelations of the atmosphere, and the terrestrial surface at determinate time and space. Nicholson [1] described that seasonal time scales define the dynamic predictability, which is explained by atmospheric fluctuations defined by internal and boundary forcing. For the internal forcing, short and medium scales are associated with mechanisms as flow instabilities, non-linear interactions, thermal and orographic forces, fluctuating zonal winds and tropical/extratropical interactions. Whereas boundary forcing could be associated to a lower boundary condition for heat and moisture fluxes related to external factors such as soil moisture, vegetation, sea-surface temperature, among others, this is also for example, soil moisture feedback on precipitation or quantifying the scales of heterogeneity in surface vegetation and soil, and their dependency with other variables as the leaf area distribution, topographical, and meteorological properties. Thus, in order to understand the different interaction process in the hydrological cycle, it will be necessary to establish the behaviour of parameters such as soil moisture at different levels of aggregation [2, 3]. For instance, on a global scale, soil moisture is important because it maintains a series of interactions with the climatic and terrestrial systems, by serving as a source of water for the atmosphere through evapotranspiration and about 60% of the amount of atmospheric water is returned to the Earth's surface in the form of precipitation [4]. As a regulator of the climate, soil moisture is linked to mass and energy cycles, affecting climatic components such as air temperature and precipitation, whose movements and disturbances presented in the atmosphere at different scales interact with the Earth surface generating heat exchange and, in consequence, supporting the stability of the air near the Earth's surface and its temperature [5]. At the basin scale, the soil moisture content is determined by the soil type and topographic configuration, and influences the partition of precipitation into infiltration and runoff and, therefore, exerts direct control over soil erosion and flooding. At the local scale, local patterns of infiltration and water flow in the soil could affect the surface water quality and groundwater [6]. This means, soil moisture is a variable that directly influences parameters such as precipitation, runoff, evapotranspiration, and infiltration since they depend on the water stored in the soil how defines the degree of modification of the water cycle parameters [5]. Under these terms, soil moisture can be understood as a relevant indicator of the alteration suffered by the climate of a given region due to the interactions with soil, vegetation and the atmosphere affecting directly plants water stress. Also, it is linked to other environmental disturbances such as solar radiation, albedo, surface temperature and water vapour gradients, which was mentioned to control the radiative fluxes between the surface and the atmosphere. This makes soil moisture to show a major complexity, since it has a synoptic condition establishing a two-way land-atmospheric interaction defining spatial patterns and temporal dynamics. Also, having information regarding soil moisture is complex due to its spatial and temporal variability; thus, having a continuous and complete database is difficult. In situ, precise measurements can be obtained but when trying to extrapolate them to a major scale, they are not reliable, which generates uncertainty when trying to use them directly to estimated parameters or in the use of hydrological models. The aim of this chapter is to analyse the database generated for SWC in a complex ecosystem using different methods under different latency and spatially in order to answer that (a) shortage periods could redefine water availability and (b) dielectric methods are a real SWC option in an ecosystem highly dynamic.

2. Soil moisture

Arnell [7] defined soil moisture as the amount of water stored in the non-saturated zone, where the soil is made of different layers or horizons (soil profile) each with different properties. These soil properties vary depending on the depth and type of rock that forms it, as well as the time at which the soil has developed and the processes that affect it. As the amount of water present in the soil layers depends on the variation of rainfall intensity and the degree of runoff or infiltration after a storm; areas with rainfall >1800 mm are considered wet, 700–1800 mm are wet-dry and <700 mm are dry. Also, loss of moisture in the soil can be as water vapour by evaporation, extraction of plant roots, transpiration or drainage in deeper layers, being the first two more significant during periods of drought [5, 8]. Within the soil, water presents a dynamic behaviour according to the potential water gradients dominated by hygroscopic and gravity for the saturated moisture content and, by capillarity under drier conditions. The hygroscopic soil moisture is defined as the amount of water that adheres to the surface of the soil particles forming a thin film; this humidity is not available for the root zone. Gravitational moisture is the amount of water that enters from the surface of the soil to the unsaturated zone in a vertical movement. Finally, capillary moisture is the amount of water available to the roots [9].

The state of soil moisture could be described in terms of the amount of water and the energy associated with the forces that hold the water in the soil. Both water content and water potential are related to a particular soil by the physical properties such as plant growth, soil temperature, chemical transport and ground water recharge. The amount of water is defined by water content, and the energy state of the water is the water potential. At this, the terrestrial water balance for a surface soil layer, which includes vegetation but not the lateral exchange between adjacent soil volumes, is expressed as:

$$\frac{dS}{dt} = P - E - R_s - R_g \tag{1}$$

where $\frac{dS}{dt}$ is the change of water content within a layer of soil, which considers soil moisture, surface water, ice and groundwater, P is precipitation, E is evapotranspiration, R_s is surface runoff and R_g is underground drainage. As soil moisture is not homogeneously distributed varying vertically and horizontally, it differs based on the soil volume being considered. Following this, the soil water content can be expressed based on its distribution in mass and volume; it is function of the apparent density. In the case of the volumetric water content, it is expressed in units of volume of water per volume of soil [$m^3\ m^{-3}$], while for the content of gravimetric water, there is a relation between the mass of water per unit mass of dry soil [$kg \cdot kg^{-1}$] [4, 10].

Soil moisture (θ) is expressed as the ratio of the total volume of soil that is wet [7]:

$$\theta = \frac{V_w}{V_T} \tag{2}$$

where V_w is the wet volume and V_T is the volume of the soil both measured in cm^3. In practice, only a fraction of the soil moisture is measured, which refers to a volume of soil. In the case of the energy balance in a soil layer, the partition of energy between soil and air is influenced by the presence and order of magnitude of soil moisture, and it could be express as:

$$\frac{dH}{dt} = R_n - \lambda E - SH - G \tag{3}$$

where $\frac{dH}{dt}$ is the change of energy within a layer of soil, including vegetation, temperature and change of phases associated with aquifers as part of the water balance; R_n is net radiation, which considers the differences between short and long wave solar energy input and output; SH is the sensitive heat flow; and G is the soil energy flow from deep layers of soil to the surface. Here, soil water potential is an expression of the energy state of water in soil and must be known or estimated to describe water fluxes. The last means the movement of water that occurs within the soil profile, between the soil and plant roots, and between the soil and the atmosphere. This movement throughout the soil is dependent on energy gradients, which includes adhesive and cohesive forces. The magnitude of the forces depends on texture and the physical-chemical properties of the soil solid matter. The differences in water potential between different soil positions cause the water to flow in it, moving from the points where the potential is greatest to those where it is least [11]. The saturated zone corresponds to the surface hydrostatic pressure that is equal to the atmospheric pressure. In the unsaturated zone, the volume occupied by the pores is filled with water and air including the area that starts at the surface of the soil and limits with the saturation zone where the water is suspended by capillary forces. As soil moisture is in the unsaturated zone, it is related to parameters such as field capacity (CC), maximum retention, percentage of permanent humidity, hygroscopic coefficient, permanent wilting point (PMP), soil tension, evapotranspiration, among others [12]. The CC is defined as the amount of water that can be retained by the soil against the outside of gravity, whereas the PMP is the amount of moisture that is not enough to stop the wilting of the vegetation.

2.1. Soil moisture measurements

There are different methodologies to measure soil water content (SWC) at the different scales: local, field, basin, and regional and global scale. Also, the transitional zone between each scale could be monitored in terms of having a better description of the condition in the area. However, one restriction related to the SWC measurement in large scales is the installation of the instruments along the study site in contact with the soil, since they need supervision and maintenance. Generally, SWC can be measured directly or indirectly. In the first case, the amount of water in the soil is determined physically, by measuring its weight as a fraction of the total soil weight by a thermalgravimetric method [13]. Some errors (bias), as well as imprecisions (larger variance), could occur when volumetric water content is calculated using an assumed bulk density or one measured elsewhere or at another time [14]. Additionally, to the direct measurement of SWC obtained, other advantages are the simple of the equipment required and that it is used as a standard method useful for the construction of calibration curves for other instruments. The main disadvantage of them is related to its destructive nature since the soil sample is removed from the field, and in consequence, the medium is destroyed and disturbs the soil profile; thus, no repetitive observations should be made. Consequently, there are very large temporary resolutions for extensive measurement networks. Also, it is a time-consuming and impractical way of measuring SWC in large scales [13, 15]. In the second case, the indirect methods estimate the humidity present in the soil by measuring another variable affected by the SWC; thus, any changes observed for this variable represent a change in SWC [15]. This type of measurement could also be sub-divided as in situ and remote methods. In in situ measurements, the instrument registered the variable affected by SWC in direct contact with the ground, whereas in the remote case, instruments are not in contact with the ground, and in fact, instruments are ported in satellite, aeroplanes, or other aerial equipment. In any case, they need to be calibrated through the generation of calibration curves using as base the gravimetric SWC.

There are several indirect in situ methods to estimate soil moisture; one of them is the volumetric. This method determines the volumetric moisture of the soil, and some examples are neutron moderation, nuclear magnetic resonance (NMR), and dielectric. The last one measures the ability of a substance to hold the charge (dielectric permittivity). The dielectric permittivity or constant of the soil determines the speed with which an electromagnetic signal is propagated within the soil. They are based on the principle of reflectometry in various domains such as time and frequency [16–19]. The dielectric constant of the soil (Ka) is dependent on the moisture content and, to a lesser degree, on the texture, temperature of the soil, bulk soil and electrical conductivity (EC). Thus, it is required to consider this dependency in order to select not only the appropriate sensor to be used, but also, those sensor properties such as geometric and electronic features [19, 20]. The value of the soil dielectric constant (Ka) is characterised by the contribution of each of its components in the soil: water (Ka $\approx$ 81), solid (Ka = 4–16) and air (Ka = 1), and it can be affected by temperature, salinity, presence of organic matter and shape and size of solid soil particles [21, 22]. These differences make the dielectric permittivity very sensitive to SWC variations [23, 24].

Time domain reflectometry (TDR) determined soil moisture by measuring the transit time of an electromagnetic pulse launched along a parallel metallic probe buried in the soil. It has been shown that the pulse travel time is proportional to the apparent dielectric constant of the soil [25]. Thus, the dissipation signal is proportional to the electrical conductivity of the soil mass; a higher content of water will provide a better propagation velocity [26]. For that water content estimation, once the instrument was calibrated, it can be related to the travel time or to the apparent dielectric permittivity (εa). The main advantages of this technology are its high accuracy, it can be automated, it provides simple measurements, and it is soil texture-, porosity-, temperature- and salinity independent [25]. For different types of soils, there is a direct relationship between the water content (θ) and the apparent dielectric constant (Ka). Some disadvantages are related to the cost of the equipment to install the sensors and to its limited applicability of the sensor in soils with conditions of high salinity or in soils with highly conductive clays [21].

Frequency domain reflectometry (FDR) provides a continuous measurement of the SWC, by means of an electromagnetic wave that is transmitted along probes and records the frequency of the reflected wave; it presents variations depending on the dielectric properties of the soil measured through the capacitance [13]. This is because the sensors work as part of a capacitor in which the water molecules are polarised and aligned in a dipolar electric field. The capacitor consists of two hollow cylindrical metal electrodes arranged coaxially but separated by several millimetres with an insulating plastic, and the use of an electronic oscillator produces a sinusoidal waveform [27]. This allow the capacitor to interact with the soil outside of the tube; thus, the capacitance measured will be affected by the soil bulk electrical permittivity and the dipoles respond to the frequency of the electric field, which can determine the capacitance that leads to know the dielectric constant and, therefore, the estimation of SWC. The relationship between the frequency of oscillation and soil water content is inverse.

3. Study case

The Calakmul Biosphere Reserve (CBR) is located at the southeast of Campeche, in the municipality of Calakmul, bordered to the east by the state of Quintana Roo and to the south by the Republic of Guatemala. The natural protected area of Calakmul is characterised by sustaining the ecosystem for different species, as well as being used as a source for water supply for animals and humans. The CBR border coordinates are 19°15′N–90°10′E, 17°45′N–90°10′E, 19°15′N–89°15′ and 17°45′N–89°15′E (**Figure 1**) [28, 29].

The CBR has a very particular climatology, edaphology and vegetation, representing a great contribution to the maintenance of the essential ecological processes, such as water and climate regimes and the ecological and evolutionary processes that determine the biodiversity of the area. Most of the land corresponds to a plateau, originated by the erosion of the limestone, little uneven, interrupted by small hills and micro-valleys. Most of the substrate is composed mainly of carbonated rock ($CaCO_3$) or limestone, and this type of substrate facilitates the filtration and underground transport of water. There are also regions that have more evaporites, rocks formed by the evaporation of marine waters ($CaSO_4$ mainly in the area), which

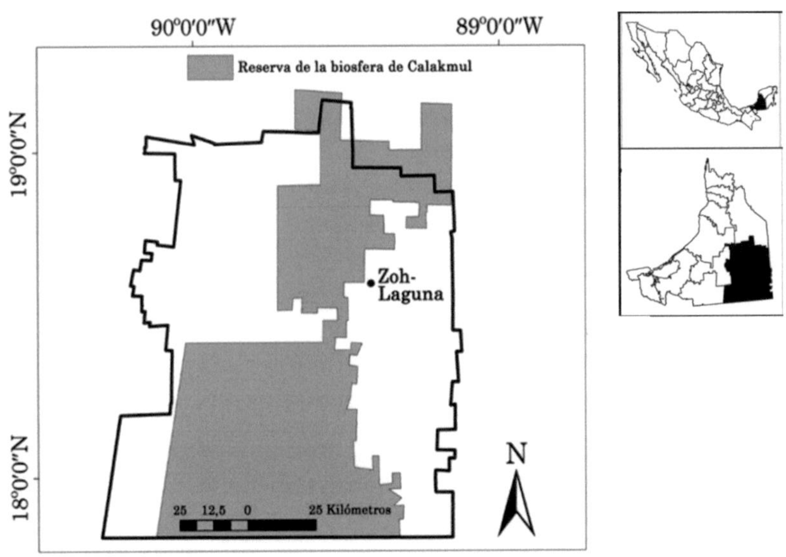

Figure 1. Location of the Calakmul Biosphere Reserve, Campeche. Source [29].

can be dissolved more easily, allowing faster erosion, and alluvium [30]. The composition of the soil allows the development of the karstic system, where water is filtered, dissolving the rock and creating underground tunnels where the liquid finally flows. These tunnels create a drainage system that feeds certain bodies of water located west and northwest of the Yucatan Peninsula, and this process favours the formation of cenotes, aguadas, wetlands, basins, caverns and springs [31].

The biodiversity contains ecosystems characterised by their great diversity, wealth and fragility. Vegetation corresponds to high jungle subperennifolia (25–50% of the trees drop their leaves), medium jungle subperennifolia, medium jungle subcaducifolia (50–75% of the trees lose their leaves), low jungle subperennifolia, savanna, aquatic vegetation and thorny scrub. One important aspect is that roots grow horizontally due to the karstic nature of the area [32–34]. The CBR is allocated in one of the hottest and wettest regions of Mexico. The climate in the CBR is warm and sub-humid with summer rainfall. The average annual precipitation in the region is 1092 mm. Rain is distributed in the months of May to October with 75% of the annual sheet, with an extension of this season until November. The months June to September are the ones observing more abundant precipitation with an average sheet from 135 to 184 mm. The dry season includes the months of December to April, during which the precipitation is less than 50 mm, and the month with the lowest precipitation is February, with an average of 33.9 mm. The average annual temperature is 24–28°C due to the vegetation that regulates it.

García et al. [35] indicated that as two slopes divide the Yucatan Peninsula: the Gulf of Mexico and the Caribbean, and the CBR is allocated in the intermedium area being subject to high scarcity. In addition, there are real water pressure in the surrounding area to use water for social development, which is manifested in the constant colonisation of the area and therefore in the opening of new crop and livestock sites. Virtually, all the rainwater infiltrates, which produces little or no runoff and the local rainfall is concentrated in small superficial storage called "aguadas," which hardly maintain the liquid until the following rainy season. Although legally human activities are restricted in the CBR to a tolerant zone, it is being severely affected by irregular human settlements that eliminate the forest to induce changes in land use.

3.1. Selection of point measured sites

The total region was analysed applying first a regular 500 × 500-m grid resulting in systematic 50 sites distributed within the whole area. Then, a zigzag statistical method was used reducing the sample to 18 sites. The priority was to allocate an aguada with or without human impact and within a town to guarantee its maintenance, thus at the field, as some of them were inaccessible, the final sites were nine. The sites were distributed as: three in the northern zone (Refugio, Flores Magón and Modesto Ángel), three in the southern zone (Carlos A. Madrazo—two sites: La Ceiba and Corosal, and Ley de Fomento) and three in the archaeological zone (Ramonal, Bonfil and Heliport) of the CBR. In the case of Carlos A Madrazo, La Ceiba aguada was only analysed since Corosal presented eutrophication.

3.2. Instruments

A monitoring station was installed in the north of the Calakmul Biosphere Reserve in the town Modesto Angel (MA) and at the South close to Ley de Fomento town was an Automatic Weather Station [36]. Soil moisture was measured using the direct gravimetric method and also continuously using indirect methods based on reflectometry: time domain (TDR) and frequency domain (FDR). Additionally, other measured variables at this station were physical characteristics of the soil, rainfall, air temperature, and relative humidity.

TDR sensors used in this study are CS616 (CS) from Campbell placed at 2.5, 5, 10, 20 and 30 cm with a latency of every 20 min. The calibration of the CS616 sensors was done according to the manufacturer (ref). In particular, these sensors use linear or quadratic equations to estimate the volumetric water content, depending on the expected range of water content and accuracy requirements. The accuracy reported for these probes is ±2.5 volumetric water content. Measurements of CS sensors are stored in a Campbell CR800 datalogger, which records the data and can then be accessed via peripheral communications using a software interface provided by the company.

FDR sensors tested were Decagon EC-5 and Diviner 2000. Decagon EC-5 (DEC) sensors measure the dielectric constant operates at 70 MHz minimising salinity and texture effects. An advantage is that they provide an accurate sensor reading in almost any soil. Factory calibrations are provided for mineral soils, potting soil and others. The design and measurement frequency allows measurement of volumetric water content (VWC) [37]. The EC-5 sensors were connected by a 3.5-mm stereo jack plug to the Generation I THHINK datalogger collecting data every 20 min [38]. Diviner 2000 [39] is a multi-sensor capacitance probe used to determine soil water content by measuring the frequency change induced by the changing permittivity of the soil permeated by the fringing fields of the capacitor sensor. The probe consists of multiple sensors located at various depths installed in specific access tubes. A high-frequency electric field is created around each sensor (sphere of influence). The sphere of influence is every 10 cm, thus readings are taken in 10 cm depth intervals in the access tube; this allows the sphere of influence for each reading to sample a separate soil horizon. Volumetric soils water measurements are done in real time and the readings are converted to soil moisture using a calibration equation. This universal calibration equation is independent of soil temperature but could be affected by salinity. One advantage is that the access tube is installed with minimum disruption to the soil profile. The accuracy level is better than 99% of the volumetric soil water content (θv) that is taken instantaneously with excellent repeatability. An access tube was allocated at each of the nine test sites into the soil to different depths until 150 cm, and in some cases just above the water table. Readings were registered every 3 days the first weeks and then every 15 days. Results were used applying the calibration equation in order to have volumetric water content and to compare with the gravimetric, TDR, and FDR (Decagon) methods.

3.3. Soil moisture measurement procedure

Field campaigns were performed in September 2012, February and August 2013, May and September-October 2014 and June 2015. Dates correspond to the rainy and dry periods, to collect representative data of each of them and in this way observe the distribution of soil moisture in different climatic regimes. The rainy season occurs between the months of June and July, until October and the dry season between December and April or May. During the visits to the study area, the physical condition of the equipment and the environment was recorded.

Fieldwork consisted of the installation of equipment, acquisition of soil moisture, vegetation, and meteorological data, which was done every fortnight, period corresponding to the data collection with the diviner sensor. Soil samples were taken for the measurement of gravimetric humidity. Soil samples were of approximately 300 g and obtained in each of the eight sites, and the following data were obtained: soil moisture and physical properties (textural fraction, bulk density, permanent wilting point (PMP), field capacity (CC), electrical conductivity and pH). Subsequently, samples of 100 g were taken every 10 cm in the soil profile to perform the gravimetric procedure and define the amount of gravimetric water content of each of the study sites. The physical characteristics of the soil samples such as bulk density, PWP and CC, electrical conductivity and pH were carried out in the National Forestry, Agriculture and Livestock Research Institute (INIFAP) and in the Soils and Plants Laboratory of the Academic Division of Agricultural Sciences of the Autonomous Juarez University of Tabasco. Once the humidity values of the indirect measurements in situ have been validated, the analysis of their spatial and temporal distribution is carried out making use of geographic information systems and other computer programs for the graphic modelling of the data.

The vertical analysis allows the visualisation of the fluctuation of soil moisture for each site, taking into account the relationship with the textural fraction of the soil. The results of this analysis permit the understanding of the mechanism of infiltration, drainage and saturation in the first meters of the soil layer. The temporal resolution to obtain one measurement varies for each technique. The highest temporal resolution can be provided by the TDR and FDR-Decagon (FDR_Dec) with one observation for every 20 min, the FDR-Diviner 2000 (FDR-Div) can record one measurement for every week, and the gravimetric method can be used for every 4 months. This indicates that one can have more frequent TDR and FDR-Dec observations than the other FDR techniques.

Once measurements with sufficient support at the local scale are obtained and the spatial and temporal stability are established, they can be scaled. Scaling up soil moisture is divided into two categories: small scale or less than 20 km^2, affected by variations in soil characteristics, heterogeneity and changes in soil cover; and regional scale, from 50 to 400 km^2, impacted by meteorological and climatological effects such as precipitation or solar radiation [3]. In this paper, a small scale is presented since the radium of influence is less than 50 km.

3.4. Soil moisture comparison

In order to estimate the accuracy between the three soil moisture methods, a comparison analysis was performed. Statistical indicators such as the coefficient of determination (R^2), the root mean square error (RMSE), relative error, mean bias error (MBE) and normalised root mean square error (NRMSE) were applied [40].

$$RMSE = \sqrt{\frac{1}{n}\sum_{t=1}^{N}(D_i - Obs_i)^2} \tag{4}$$

where subscripts D_i is the output of the devices (FDR and TDR readings) and Obs_i is the observed gravimetric soil moisture. *RMSE* minimum value is zero under the hypothetical situation that the model is capable of perfect (long-term) readings of the system, and there are no data errors being small values desirable. Mean bias error (MBE) measures the average magnitude of the errors in a set of readings. It is the average over the test sample of the absolute differences between prediction and actual observations having the differences an equal weight. According to [41] an acceptable value for volumetric soil moisture is 0.04 $m^3\ m^{-3}$.

4. Results and discussion

The results present the climatic variations of the studied area that make it a complex system to analyse the variability of soil moisture in the sampling sites. The test of the TDR and FDR systems can offer more than the evaluation of the accuracy of each system if they are used as well as a complementary study. Also, a datalogger was tested to work in a complex environment in order to guarantee a constant data register in order to monitor water requirements and its supply for the different uses in the area.

4.1. Climatic data

As it was mentioned, the climate that predominates in the region is warm humid and warm sub-humid, with an average maximum temperature of 36°C during the months of May and June and average minimum temperature of 18°C during January. However, in some occasions during August, the so-called *dog days* is presented, which is the year period where heat

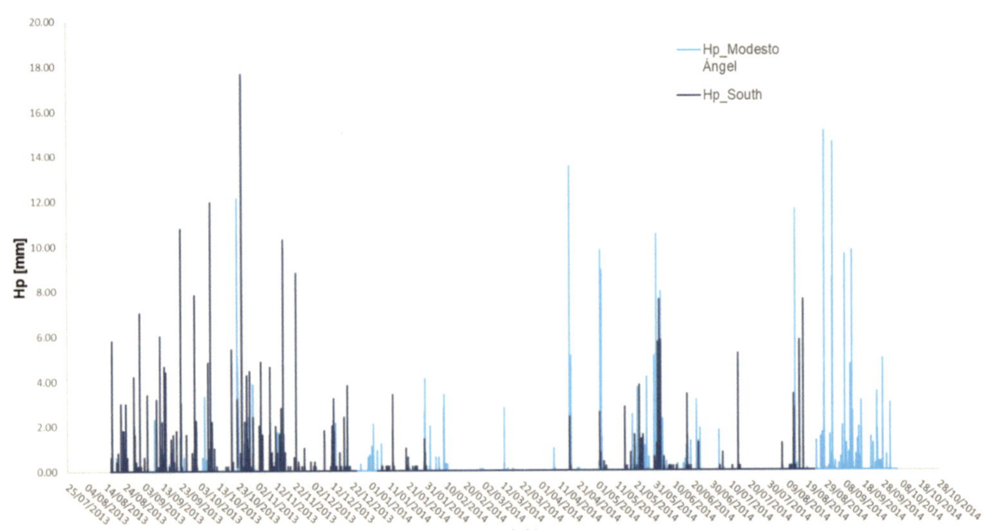

Figure 2. Distribution of head of precipitation recorded from August 2013 to October 2014 in the Modesto Ángel station and the Automatic Weather Station (EMA)-CONANP station at the north and south of the study of the CBR, respectively.

is severe and drought is taking place. According to [30], the rainy season occurs during the middle of summer until autumn, although it was observed in **Figure 2** that occurs from July to February. During the winter, the tropical storms have influence on the generation of precipitations and also they cause a decrease of the temperature until reaching 10°C [30].

Figure 2 shows the rainfall distribution in the study area. Although the north area seems more affected by the winter rainfall, there is more rain in the south part of the CBR. Besides, in the dry season from April to July, there was some rainfall at the north area. This variability favoured the presence of humidity maintaining the diversity of vegetation species in both areas. The archaeological zone is considered at the south.

4.2. Soil characteristics

The characterisation of the soil is based on the measurement of its texture on the surface and some physical parameters such as electric conductivity, pH, soil moisture, % of saturation, field capacity (CC) and permanent wilt point (PWP) (see **Table 1**). CC is the largest amount of water that this type of soil will retain under conditions of complete humidity $CC = (\%clay) * a + (\%sil) * b + (\%sand) * c$, and the PWP is the minimum water content where the plants usually die; for each of the sampling soils $PWP = \frac{CC}{1.84}$, the coefficients a, b and c are determined for each region and type of floor, and in this case, the coefficients used are $a = 0.555$, $b = 0.187$ and $c = 0.027$ [42].

Modesto Ángel where the three techniques were implemented has a more constant type of soil: the first 5 cm is a sandy soil, from 10 to 80 cm is Frank and more than 90 cm is loamy clay. In all the other sites, soil type varies as the depth increases.

Site	Sand	Clay	Silt	Type of soil	% Humidity	% Saturation	CC	PWP	pH	Electric conductivity
Flores Magón	53.28	15.08	31.64	Sandy loam	37.04	60.0	45.0	23.6	7.38	0.291
Refugio	12.00	58.56	29.44	Clayish	37.04	104.0	78.0	41.0	6.42	0.251
Modesto Ángel	31.84	31.80	36.36	Frank	30.77	102.2	76.7	40.2	7.23	0.361
Ley Fomento	26.20	42.16	31.64	Clay	38.34	107.0	80.3	42.1	6.57	0.551
La Ceiba	31.84	27.80	40.36	Frank	39.13	120.0	90.0	47.3	7.93	1.101
Ramonal	74.40	7.24	18.36	Sandy loam	35.71	65.8	49.4	25.9	6.80	0.797
Bonfil	30.76	31.24	38.00	loamy sandy	37.93	125.0	93.8	49.2	6.84	0.564
Helipuerto	32.20	25.80	42.00	loamy sandy	32.38	120.2	90.2	47.3	5.43	0.024

Table 1. Soil characteristics measured at 10 cm depth in the sites.

4.3. Soil moisture analysis

Soil moisture instruments tested report changes in time or frequency related indirectly to the dielectric permittivity to the volumetric water contents. Results are presented per site and per type of technique, and this means that TDR were analysed with Campbell sensors (TDR_Campbell), FDR using the Decagon sensors (FDR_Dec) and FDR using Diviner 2000 (FDR_Div). A specific analysis for the field conditions under the different operating sensors was not necessary since there were the same conditions at the sites (Modesto Ángel with three methods and the other two sites). Individual calibrations per depth offer equations that improve the sensor performance. Then all together were compared in order to know the sensitivity of each one.

4.3.1. FDR and TDR devices at the Modesto Ángel station

In the Modesto Ángel station, the three devices to measure soil moisture were installed: TDR_CS, FDR_Dec and FDR_Div. TDR_CS calibration process includes a first analysis using data collected from the datalogger with the default equations of the device. Both lineal and quadratic equations were tested founding that the lineal equation offered better results than the quadratic one with RE of 0.22 and 0.41, respectively. Secondly, the calibration using the gravimetric data measured at the field was done in terms of volumetric water moisture, θgv. The gravimetric measurements were 10 samples per site from 2012 to 2013 years, at 2.4, 5, 10, 20 30 cm depth. Six more gravimetric samples per site were measured during 2014–2015 to confirm the reliability of the calibration for a different weather, soil and vegetation conditions. The TDR_CS provides a R^2 of 0.96, with RMSE, MBE and RE values of 0.101 $cm^3 cm^{-3}$, 0.107 $cm^3 cm^{-3}$ and 0.12, respectively. The major deviation was observed at the 2.5 and 5 cm, this is because the place where the sensors were installed was not disturbed but the place where the sample was taken, even if it was close to the area of the station, was more susceptible to the surface soil conditions at this specific time. In the case of the FDR_Dec, R^2 once calibrated the readings was 0.90 with RMSE, MBE and RE values of 0.18 $cm^3 cm^{-3}$, 0.407 $cm^3 cm^{-3}$ and 0.083, respectively. For FDR_Div, results showed that the first 30 cm have the same texture; thus, a calibration equation was obtained for it with an R^2 of 0.86, RMSE 0.086 $cm^3 cm^{-3}$, MBE 0.079 $cm^3 cm^{-3}$ and RE 0.069. FDR_Div was the best device, but it is important to mention that only three depths were tested at 10, 20 and 30 cm. Between TDR_CS and FDR_Dec, it is quite difficult to analyse since results are not conclusive, but TDR_CS could be expected to provide a more consistent value. This is because Decagon devices demonstrated to be more sensible to the weather (see temperature and precipitation results) and to soil conditions at the time of the sampling. However, the RE is major for TDR_CS. Despite the previous results, the calibration equations were applied to 400 daily θs records with a latency of 20 min for the 2012–2015 period for both TDR_CS and FDR_Dec. **Figure 3** plots the behaviour of the devices at a 10 cm depth for 2013–2014.

As one can observe, there is a good agreement between TDR_CS and FDR_Dec, following a similar pattern taking into account the accuracy of each device. However, there is a major response of the Decagon sensor when rain is presented, being evident an increase in some

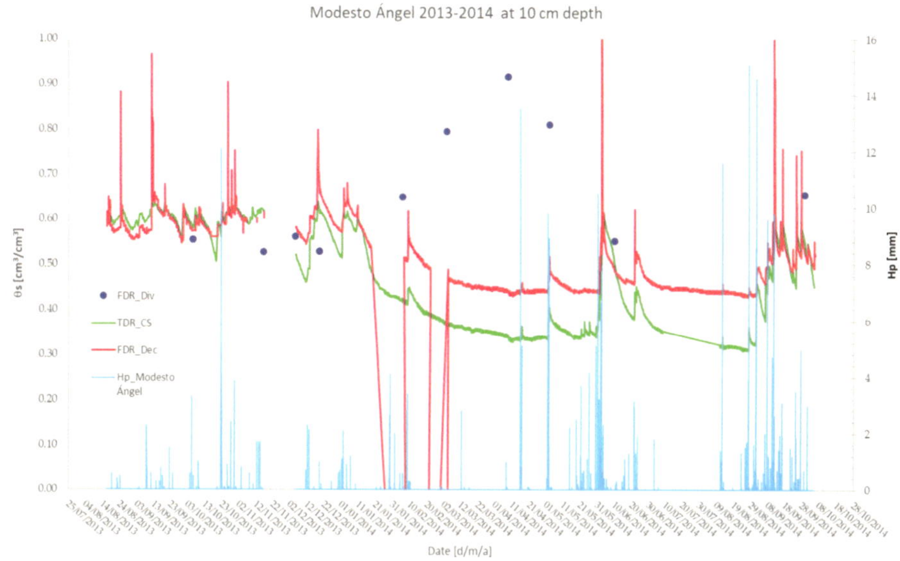

Figure 3. 10 cm depth soil moisture values from July 2013 to October 2014 at the Modesto Ángel station.

cases until reaching almost 1.0 of water content. The maximum difference perceived from January to August 2014 was close to 30%. Looking at the FDR system, it is a major agreement for both Decagon and Diviner 2000, although during the dry period, FDR_Div overestimated more than 50% the soil moisture. The gravimetric measurements for October 31, 2013 and May 05, 2014 were also included in the plot demonstrating that TDR_CS is closer than FDR_Dec.

The life of the experimental THINNK datalogger without change of battery was from 2013 to 2015. In the case of the Campbell datalogger, it was required to change the battery since the extreme conditions at the field lowered its energy every 6 or 8 months. Also, it was necessary to protect the battery of the Campbell datalogger, whereas in the THINNK, one could be attached to a tree without more protection.

4.3.2. *FDR diviner for the other sampling sites*

The Diviner 2000 allows monitoring different areas once the accessed pipe in each site was installed. Thus, eight aguadas were monitored; Carlos A Madrazo was analysed at the La Ceiba site only. More than 30 readings were registered in the period of 2012–2015, as well as several gravimetric analyses were performed a less two per year. As soil moisture is function of the texture along the profile, this implicates different water aggregation and, in consequence, a different behaviour. For that, along the profile, one could have more than one calibration equations in order to represent what actually happened to soil moisture in the profile. For each site, a graph was developed as shown in **Figure 4** for the Modesto Ángel site with 120 cm depth. For this place, three equations were established:

$$(a)\ 10\text{–}30\ \text{cm}: y = 425.6\,x^{-0.597}; R^2 = 0.93 \tag{5}$$

$$(b)\ 40\text{–}80\ \text{cm}: y = 37.33\,x^{0.123}; R^2 = 0.02 \tag{6}$$

$$(c)\ 90\text{–}110\ \text{cm}: y = 136.8\,x^{0.308}; R^2 = 0.35 \tag{7}$$

The equation proposals agree with the findings of [43] who defined two equations according to the texture: one group for fields with heavier soils where clay content was >40% and other group with coarser textured fields with clay content <40%. Looking at Eq. (6) for the second

Depth	Type of soil	CC	PWP
10 cm	Clayish	76.7	40.2
20 cm	Clayish	75	39.4
30 cm	Clayish	75	39.4
40 cm	Frank	87.2	45.8
50 cm	Clayish	78.2	41
60 cm	Clayish	77.7	40.8
70 cm	Clayish	77.1	40.5
80 cm	Clayish	79.2	41.6
90 cm	clay-silt-loam	83.6	43.9
100 cm	Clay	84	44.1
110 cm	Clayish	75	39.4
120 cm	Clayish	78.3	41.1

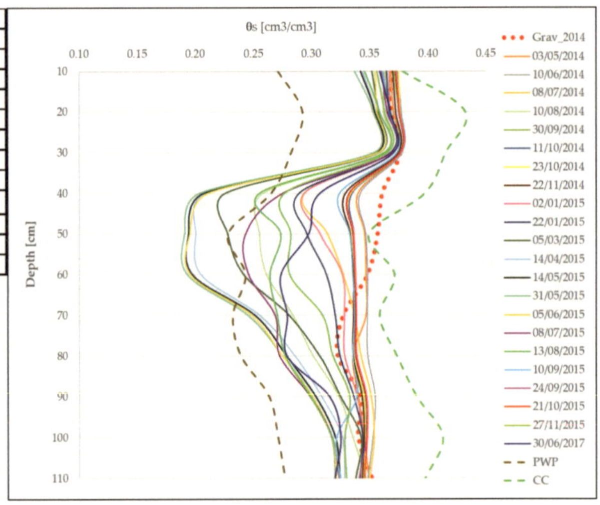

Figure 4. The texture and field capacity (CC) and permanent wilt point (PWP) through the profile of the Modesto Ángel site.

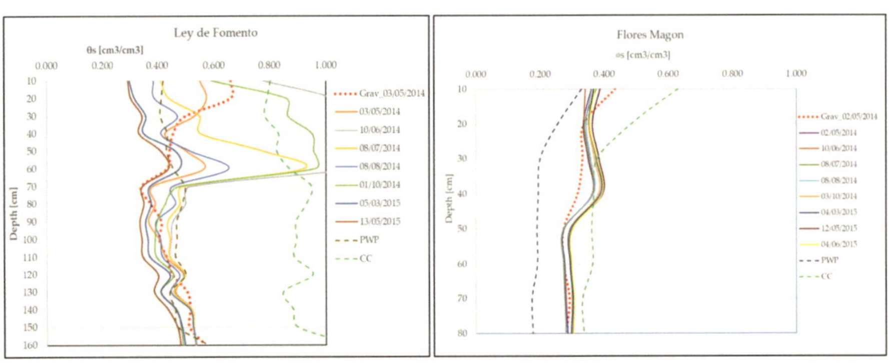

Figure 5. Soil moisture profile for Ley the Fomento and Flores Magón sites.

group shows a very poor correlation, there is a tendency towards the PWP, and in some cases, it seems that water was not available from the 40–80 cm depth. For these specific cases, FDR_Div values correspond to late March to mid-June of 2015 where a drought also took place. In addition, the drop between these depths was evident in the gravimetric measurements, although it was not in such an accentuated way. This could be interpreted as a combined effect of soil properties in these strata, the horizontal development of vegetation root systems and the dielectrical methods characteristics in that range of soil moisture. However, this requires a further research.

One interesting factor resulted from the analysis at the sites of Flores Magón and Ley de Fomento was that both FDR_Div and the gravimetric carried out the same day with less than 30 min of difference had not been similitude within the first 50 cm depth (**Figure 5**). In these cases, it was assumed that the disturbing at the moment of sampling in the gravimetric method was significant to obtain different values. Also, in these sites, the aguadas were extended and the types of soil were clayish in Ley de Fomento, and frank and clay in Flores Magón, conditions that change the water content in the soil.

4.4. Water availability

Although one knows from [35] that the approximated number of the aguadas in the CBR was 1353 and the majority was in the North with 868 and 485 in the South, with an approximate density of 1.87 aguadas per hectare. One does not know the rate of increment in the number of aguadas, in particular for the artificial ones since they are the only source of water during the dry season. Also, the area of the aguadas is variable from some small as one observed in La Ceiba at Carlos A Madrazo to many hectares such as Flores Magón. In terms of estimation of the water availability, it can be considered that the aguadas are generically of the same size; thus, the contribution area (basin) of each aguada can be calculated within the CBR as the relation between the total area of 724,000 ha and the total aguadas (1353), so here are 535 ha of contribution per aguada.

To estimate the water availability based on the real soil moisture during the year, it is necessary to review the distribution in the horizontal line at the different profile. Thus still, there is work to do. Until now, the work done has established the variation along the profile finding some constant values after the 50 cm depth that could be related to the water table that keep water in some aguadas during the year although at its minimum value, in particular those less impacted by humans.

One already knows the average rainfall, and the evapotranspiration could also be estimated moreover if one consider that practically all the water infiltrates (probably 92.7%), which results in little runoff (7.3%) that is concentrated in natural aguadas. However, these data are not enough since soil moisture needs to be considered as the water storage capacity that can be removed by evapotranspiration is the function of the type of vegetation and the depth of the root zone. But, in the study area, root depths are quite smaller growing horizontally rather than vertically. This demands a major study in the horizontal line in order to compute reliable water balances.

5. Conclusions

Over the studied period, the soil moisture values at the soil profile varied in all the sites sampled. There are some similitudes grouping the areas in those sites in the north, south and in the archaeological zone. However, there are not only different soil conditions, but also vegetation, climate and topography differences. Vegetation varies from low to middle jungle with roots growing horizontally. Climate varies dividing the area in quadrants with important temperature and precipitation variations.

Soil moisture result using TDR_CS and FDR_Dec sensor in a datalogger systems was more similar after 10 cm depth. At 2.5 and 5 cm, there is a high variability mainly associated to the actual conditions in the soil surface. However, the seasonal effect under soil moisture demonstrated that FDR_Dec was very sensitive in periods of high rainfall and overestimated soil moisture during the dry season. Something similar happened using the FDR_Div sensors, which overestimated significantly in the dry season. The best result was provided by the CS616 Campbell sensors for over the whole study period. Looking at the soil moisture values along the depth profile, it was effectively probed that when it has a heterogeneous soil in the unsaturated zone, texture is determinant. If one did not consider these, a wrong calibration could be obtained and soil moisture values would have nonsense. Another important aspect to be considered is the possible disturbance of the soil at the moment of the sampling for the gravimetric method. Also, raw values from the different devices need to be calibrated; otherwise, any soil moisture value could be obtained. Even if the calibration was not appropriate under dry conditions, it is necessary to consider the karstic nature of the soil in the area; thus, infiltration could occur at different deep levels.

Acknowledgements

Authors thank to Kevin Vargas for the work done during his stage as a scholarship holder of the Instituto de Ingeniería, UNAM and also, to Alejandro Monsivasis Huertero researcher at the National Polytechnic Institute. This research was originally founded by the Instituto de Ingeniería, UNAM with the 1315 and 2335 internal projects.

Author details

Judith Guadalupe Ramos Hernández[1*], Jesus Gracia-Sánchez[1],
Tania Patricia Rodríguez-Martínez[1] and José Adalberto Zuñiga-Morales[2]

*Address all correspondence to: jramosh@iingen.unam.mx

1 Instituto de Ingeniería, Universidad Nacional Autónoma de México, Av. Universidad, Coyoacán, Mexico City, Mexico

2 Reserva de la Biosfera Calakmul, Calle Puerto Rico s/n, X-Pujil, Calakmul, Campeche, Mexico

References

[1] Nicholson SE. Land surface atmosphere interaction physical processes and surface changes and their impact. Progress in Physical Geography: Earth and Environment. 1988;**1**:36-65. DOI: 10.1177/030913338801200102

[2] Brocca L, Melone F, Moramarco T, Morbidelli R. Spatial-temporal variability of soil moisture and its estimation across scales. Water Resources Research. 2010;**46**:W02516. DOI: 10.1029/2009WR008016

[3] Entekhabi D, Rodriguez-Iturbe I. Analytical framework for the characterization of the space-time variability of soil moisture. Advances in Water Resources. 1994;**17**:35-45. DOI: 10.1016/0309-1708(94)90022-1

[4] Seneviratne SI, Corti T, Davin EL, Hirschi M, Jaeger EB, Lehner I, et al. Investigating soil moisture–climate interactions in a changing climate: A review. Earth-Science Reviews. 2010;**99**:25-161. DOI: 10.1016/j.earscire.2010.02.004

[5] Njoku EG, Entekhabi D. Passive microwave remote sensing of soil moisture. Journal of Hydrology. 1996;**184**:101-129. DOI: http://citeseerx.ist.psu.edu/viewdoc/download?doi=10.1.1.26.3857&rep=rep1&type=pdf

[6] Huisman JA, Sperl C, Bouten W, Verstraten JM. Soil water content measurements at different scales: Accuracy of time domain reflectometry and ground penetrating radar. Journal of Hydrology. 2001;**245**:48-58. DOI: 10.1016/S0022-1694(01)00336-5

[7] Arnell N, editor. Hydrology and Global Environmental Change: Understanding Global Environmental Change. New York: Person Education Limited; 2002. 337 p. ISBN-13: 978-0582369849, ISBN-10: 0582369843

[8] Osman K, editor. Soils: Principles, Properties and Management. Vol. 271. Dordrecht, The Netherlands: Springer Science + Bussines Media; 2013. DOI: 10.1111/sum.12053

[9] Monsalve G, editor. Hidrología en la ingeniería. Bogotá: Escuela Colombiana de Ingeniería, Alfomega; 1999. 360 p. ISBN: 9789701504048

[10] Bilskie J. Soil water status: Content and potential. App. Note: 2S-I. Logan, UT: Campbell Scientific, Inc [Internet]; 2001. Available from: https://s.campbellsci.com/documents/us/technical-papers/soilh20c.pdf [Accessed: January 21, 2018]

[11] Hillel D, editor. Environmental Soil Physics. San Diego, CA: Academic Press; 1998. 771 pp. ISBN: 9780123485250

[12] Behari J, editor. Microwave Dielectric Behaviour of Wet Soils. New York: Springer; 2005. 163 p. ISBN 978-1-4020-3288-2

[13] Muñoz-Carpena R. Field devices for monitoring soil water content. Extension Bul. 343. Florida: UF/IFAS Department of Agriculture, University of Florida; 2004. 15 p. DOI: http://edis.ifas.ufl.edu/AE266

[14] Evett SR, Heng LK, Moutonnet P, Nguyen ML, editors. Field Estimation of Soil Water Content: A Practical Guide to Methods, Instrumentation, and Sensor Technology. Vienna, Austria: IAEA-TCS-30. Intl. Atomic Energy Agency; 2008. p. 131. ISSN 1018-5518

[15] Bittelli M. Measuring soil water content: A review. HortTechnology. 2011;**21**(3):293-300. DOI: horttech01785 293.300

[16] Evett SR, Parkin GW. The continuing maturation of technology and theory. Vadose Zone Journal. 2005;**4**:986-991. DOI: 10.2136/vzj2005.0099

[17] Gardner CMK, Dean TJ, Cooper JD. Soil moisture measurement with a high frequency capacitance sensor. Journal of Agricultural Engineering Research. 1998;**71**:395-403. DOI: 10.1006/jaer.1998.0338

[18] Robinson DA, Gardner CMK, Cooper JD. Measurement of relative permittivity in sandy soils using TDR, capacitance and theta probes: Comparison, including the effects of bulk soil electrical conductivity. Journal of Hydrology. 1999;**223**:198-211. DOI: 10.1016/S0022-1694(99)00121-3

[19] Robinson DA, Jones SB, Wraith JA, Or D, Friedmana SP. A review of advances in dielectric and electrical conductivity measurement in soils using time domain reflectometry. Vadose Zone Journal. 2003;**2**:444-475

[20] Bogena HR, Huisman JA, Oberdörster C, Vereecken H. Evaluation of a low-cost soil water content sensor for wireless network applications. Journal of Hydrology. 2007;**344**:32-42. DOI: 10.1016/j.jhydrol.2007.06.032

[21] Jones Meng L, Quiring SM. A comparison of soil moisture models using soil climate analysis network observations. American Meteorological Society. 2008;**9**:641-659. DOI: 10.1175/2008JHM916.1

[22] Wraith JM, Or D. Temperature effects on soil bulk dielectric permittivity measured by time domain reflectometry: Experimental evidence and hypothesis development. Water Resources Research. 1999;**35**:361-369. DOI: 10.1029/1998WR900006

[23] Gerhards H, Wollschläger U, Yu Q, Schiwek P, Pan X, Roth K. Continuous and simultaneous measurement of reflector depth and average soil-water content with multichannel ground-penetrating radar. Geophysics. 2008;**73**:15-23. DOI: 10.1190/1.2943669

[24] Huisman JA, Hubbard SS, Redman JD, Annan AP. Measuring soil water content with ground penetrating radar: A review. Vadose Zone Journal. 2003;**2**:476-491. DOI: 10.2113/2.4.476

[25] Topp GC, Davis JL, Annan AP. Electromagnetic determination of soil water content: Measurements in coaxial transmission lines. Water Resources Research. 1980;**16**(3):574-582. DOI: 10.1029/WR016i003p00574

[26] Martínez Fernández J, Ceballos Barbancho A. Design and Validation of a TDR probe for Measuring Soil Moisture. In: López JJ, Quemada M, editors. Temas de Investigación en

Zona no Saturada. Department of Geography, University of Salamanca, Spain; 2001. 1-7 p. DOI: http://docplayer.es/1326579-Diseno-y-validacion-de-una-sonda-tdr-parala-medicion-de-la-humedad-del-suelo-martinez-fernandez-j-y-a-ceballos-barbancho.html

[27] Evett S, Cepuder P. Capacitance sensors for use in access tubes. In: Evett SR, Heng LK, Moutonnet P, Nguyen ML, editors. Field Estimation of Soil Water Content: A Practical Guide to Methods, Instrumentation and Sensor Technology. Training Course Series 30. Vienna: IAEA; 2005. pp. 73-90. ISSN 1018-5518

[28] CONANP, editor. Conservation and management program. Calakmul Biosphere Reserve. Teaching Series. Mexico: CONANP; 2014

[29] Hurtado R, O'Farril D, Andrade M, Padilla A, Sosa L. Las aguadas de Calakmul: Reservorios de vida silvestre y de la riqueza natural de México. Mexico: CONABIO; 2010. pp. 1-6. DOI: https://www.researchgate.net/profile/Jorge_Lopez-Calderon/publication/257931284_Lopez-Calderon_et_al_2010_Biodiversitas_93_vol_completo/links/0deec526482910ad15000000/Lopez-Calderon-et-al-2010-Biodiversitas-93-vol-completo.pdf

[30] Semarnat, editor. Programa de Manejo de la Reserva de la Biosfera de Calakmul, México. Mexico: Instituto Nacional de Ecología-SEMARNAT; 1999

[31] García-Gil G, Palacio JL, Ortiz MA. Reconocimiento geomorfológico e hidrográfico de la Reserva de la Biosfera Calakmul, México. Investigaciones Geográficas. 2002;**48**:7-23. DOI: http://www.scielo.org.mx/scielo.php?script=sci_arttext&pid=S0188-46112002000200002

[32] ITESM, editor. Plan estratégico de desarrollo integral del estado de 2000-2025, Entornos, Problemática y Estructura Económica de Quintana Roo. In: Entorno Geográfico. Gobierno del Quintana Roo; 2000. Cap. II. Quintana Roo, Mexico. DOI: https://es.scribd.com/document/178448588/Plan-Estrategico-de-Desarollo-Intergal-Del-Estado-de-Qr-2000-2025-Tomo1

[33] Galindo-Leal C, editor. La Gran Región de Calakmul, Campeche: Prioridades biológicas de conservación y propuesta de modificación de la Reserva de la Biosfera: Center for Conservation Biology, Stanford. Mexico: World Wildlife Found; 1999. 40 p

[34] Carabias LJ, Provencio E, De la Maza Elvira J, de la Gala Méndez JB R, editors. Programa de manejo de la reserva de la biosfera Calakmul. 1ra Edicion ed. Mexico: Instituto Nacional de Ecología; 1999. 277 p. DOI: http://www.paot.mx/centro/ine-semarnat/anp/AN08.pdf

[35] García GG, Palacio PJL, Ortiz PMA. Reconocimiento geomorfológico e hidrográfico de la reserva de la biosfera Calakmul, Mexico. Mexico: Investigaciones Geográficas. 2002;**48**:7-23

[36] SEMARNAT-CONANP. Logros 2015. 1ra ed. Mexico: Secretaría de Medio Ambiente y Recursos Naturales/Comisión Nacional de Áreas Naturales Protegidas. 97 p. DOI: https://www.gob.mx/cms/uploads/attachment/file/242893/LOGROS_2015_web.pdf

[37] Decagon. EC-5 Lowest cost volumetric water content sensor—Small area of influence [Internet]. Available from: http://www.decagon.com/products/soils/volumetric-water-content-sensors/ec-5-soil-moisture-small-area-of-influence/ [Accessed: May 30, 2012]

[38] THHINK. THHINK Soil Moisture Sensing and Data Logger System [Internet]. Available from: http://www.thhink.com/ [Accessed: May 30, 2012]

[39] SENTEK. Portable Soil Water Monitoring Solution [Internet]. Available from: http://www.sentek.com.au/products/portable.asp [Accessed: February 15, 1995]

[40] Ojo ERT, Bullock PR, Fitzmaurice J. Field performance of five soil moisture instruments in heavy clay soils. Soil Science Society of America Journal. 2015;**79**:20-29. DOI: 10.2136/sssaj2014.06.0250

[41] Entekhabi D et al. The soil moisture active passive (SMAP) mission. In: Proceedings of the IEEE 98.5. 2010. pp. 704-716. DOI: http://hdl.handle.net/1721.1/60043

[42] Cisneros Almazan R, editor. Notes from the Field of Irrigation and Drainage. Centre for Research and Graduate Studies. San Luis Potosi, Mexico: Agrogeodesica Area Engineering Faculty of the Autonomous University of San Luis Potosi; 2003

[43] Rowlandson TL, Berg AA, Bullock PR, Ojo ER, McNairn H, Wiseman G, et al. Evaluation of several calibration procedures for a portable soil moisture sensor. Journal of Hydrology. 2013;**498**:335-344. DOI: 10.1016/j.jhydrol.2013.05.021

Section 3

Landscape Water

Chapter 5

Water Replenishment in Agricultural Soils: Dissemination of the IrrigaPot Technology

Lucieta G. Martorano, Araya A. Berhe,
José Reinaldo da Silva Cabral de Moraes,
Ayllan Rayanne da Silva Lima,
Douglas Cavalcante Costa,
Aline Michelle da Silva Barbosa and
Marcelo Coelho Marques

Additional information is available at the end of the chapter

http://dx.doi.org/10.5772/intechopen.80605

Abstract

The challenge confronted by farmers during prolonged periods of soil water stress is to guarantee the restoration of water and maintain the productivity of agricultural crops. Even in regions such as Amazon, the variability in the precipitation regime should be considered in agricultural planning. There are regions in which 80% of annual rainfall is concentrated between December and June. It is exactly during this period of low rainfall that small-scale family-based farmers need technological assistance to guarantee that their crops remain irrigated in order to maintain their income in this rural environment. The IrrigaPot arises as an alternative that is able to access rainfall that has been stored since the rainy season and provide it to plants when the soil is dry. The pots are maintained full with 20 liters of water, and through capillary action the soil maintains them constantly humid. This technology does not require specific knowledge with respect to irrigation regimes and is necessary for the farmer to dedicate his time to replacing water. The technology is totally automated through a simple system using a float, tubes, and connectors that connect a rubber hose to the lids of the pots buried in the soil.

Keywords: sustainability indicators, water security, blue water footprint, agricultural productivity

1. Introduction

Small-scale family-based farmers oftentimes suffer considerable decreases in agricultural production during prolonged periods of soil water stress. Irrigation technology using clay pots buried in the soil is promoted as being an effective, accessible, and environmentally sustainable option for small-scale family-based farmers to be able to cultivate fruit trees and vegetables, and that can promote food security [1, 2]. Irrigation technology using clay pots buried in the soil has been used in important agricultural regions in the world [1, 3, 4] such as Burkina Faso, Zambia, the USA and Pakistan. Clay pots buried in the soil have been shown to be effective in the cultivation of fruit trees and in reforestation projects [1, 3]. This technique minimizes water losses due to evapotranspiration and soil drainage during irrigation in rural areas [5], improves seed germination and crop establishment [3], thus reducing crop loss and financial loss to farmers [1].

This technology aims to provide solutions that are able to supply a crop's water needs during long dry periods [6], especially in rural areas that require irrigation to guarantee agricultural production [7]. Adopting a holistic vision of water security in regions that have an increasing demand for water in order to produce food, this technology presents indicators that point to sustainability for food security as well as for the responsible use of water resources. Therefore, locally accessible innovations that improve the efficiency of irrigation systems are necessary in order to minimize undesirable losses due to evapotranspiration and soil drainage. Such practices aim to mitigate impacts on current analyses being conducted to study the climate as well as scenarios of global climate change. Quantification of evapotranspiration rates is fundamental to the evaluation of environmental sustainability indicators. In this context, the objective of this project is to strengthen research activities and share knowledge of technology that is low-cost and that has a small water footprint that uses rainfall water for hydrologic replenishment in soils agricultural systems.

There is a great need to increase research and extension actions that make viable the diffusion of the technology of the use of rainfall water to fill clay pots buried in the soil to maintain the production of agricultural crops during prolonged periods of soil water stress under actual climate conditions and those of future scenarios influenced by climate change, and to disseminate the results in order to amplify the adoption of this technology. In the course of using the technique, new strategies of low-cost irrigation can be adapted to different production systems and also in urban environments to increase sustainability in green spaces such as parks, public squares, schools and community gardens.

Managing irrigation water is among the critical issues to address food insecurity under the changing climate. Rainfall variability has been reported to significantly impact the economies of many countries as natural rainfall is the major source of water for agriculture. Clay pot technology has been proven to significantly improve crop water productivity in dry land areas but has not been promoted or used due to the lack of a suitable crop-specific standard design. In this context, the objective of this publication is to strengthen research activities and share knowledge of technology that is low-cost, and that has a small water footprint that uses rainfall water for hydrologic replenishment in soils in agricultural systems.

2. Material and methods: low-cost and climate-smart irrigation technology

The experiment was carried out in northern Ethiopia and the results from this field work served as the data for several theses done at the University of Mekelle. The water seeps out through the micro-pores of the clay pots with relatively slow flow and larger surface wetting time, and thus promotes a greater area of coverage around the roots of plants. Contrarily, perforated clay pots leak water faster through the macro- and micro-pores and have relatively shorter wetting time and smaller area coverage.

On the other hand, the difference between perforated bars and round ones was simply the shapes of the pots which has to do with the area of coverage along the rows of the Swiss chard plant. Round types of pots were not as suitable as bar types (of the same volume) for rows of Swiss chard crops due to their wetting area coverage along the two sides of the bar.

Therefore, among the tested clay pot designs, the bar-shaped perforated clay pot designs were evaluated as best in terms of biomass yield and economic water-use efficiency. The water-use efficiency, economic aspects, and biomass for the perforated bar clay pot design were better than that of the bucket irrigation system. The other advantages of perforated bar clay pots over the bucket type is that the water source is inside the soil thus evaporation is almost zero and there is also less probability of occurrence of leaf disease due to wetting, and this ultimately improves the biomass and water-use efficiency.

2.1. Water-use evaluation

2.1.1. Measuring evapotranspiration using the surface renewal technique

$$LE = Rn - G - H \tag{1}$$

where, LE is latent heat flux, Rn is net radiation, G is soil heat flux, and H is sensible heat flux.

2.2. Calculation of surface renewal (SR) and measuring Rn and G

The calculation of SR is done using Eq. (1). This is a residual energy balance equation. The net radiometer and soil heat flux plate data will be measured every 5 minutes and then averaged and recorded at the end of each 30 minutes. Soil temperature data will be recorded at the end of each 30 minutes, and the change in soil heat storage (dS) above the heat flux plates can be computed as in Eq. (2):

$$dS = VC \times ((T\ final - T\ initial)/1800) \times D \quad (2)$$

where VC = apparent volumetric heat capacity of the soil; T final and T initial = final and initial temperatures for a 30 minute period, and D = 0.04 m = depth of the heat flux plate. The value 1800 is the number of seconds for each 30 minutes. The VC is calculated as the product of the apparent soil density and the specific heat. The soil heat flux density at 0.04 m depth (G0) is calculated as the mean of the two heat flux plate measurements. Then the soil heat flux density at the surface (G) was calculated as:

$$G = dS + Go \quad (3)$$

2.3. Calculating surface renewal sensible heat flux

Temperature data was collected at a frequency of 4 Hz and the time lags of r = 0.25 and 0.5 s were used in a structure function to determine the temperature ramp amplitude (Ar) and inverse ramp frequency (D + S) as described [8]. The uncalibrated sensible heat flux density (H′).

$$H' = q \times Cp \times ((Ar)/D + S)) \times Z \quad (4)$$

where q is air density (kg m^{-3}); Cp = specific heat at constant pressure (J kg^{-1} K^{-1}) of the air; and Z is measurement height (m). A calibration factor (f) was used to account for uneven heating below the temperature measurement height and other potential issues [9] and to convert the uncalibrated H′ to the actual sensible heat flux density.

$$H = f \times H' \quad (5)$$

The 'f' values depend on the thermocouple size, sampling frequency, height above the ground, and the underlying vegetation [8–10]. A calibration factor was be determined using a linear regression of sonic anemometer H readings versus H′ data collected over a one-week period on the site.

Reference evapotranspiration (ETo) was based on FAO-penman Montheith [11, 12]. Determination of crop coefficient (kc) and actual and maximum evapotranspiration (ETa and ETc): from Eq. (1), LE can be related to ETc or ETa;

$$kc = ETc/ETo \quad (6)$$

where ETc, is average crop maximum evapotranspiration per week; ETo is the average weekly reference evapotranspiration.

$$Ks = ETa/ETc \tag{7}$$

where ETa is actual evapotranspiration and ks is the stress cofficient.

$$ETa = ks \times ETc \text{ or } ETo \times kc \times ks \tag{8}$$

2.4. Water-use evaluation in Africa

Water applied at each site was evaluated based on water held in the soil and data from production and harvest. Water-use efficiency is used as an important parameter to evaluate the performance of this technology. The water-use efficiency is calculated using harvest yield (kg) per m^3 of water applied to the crop. Water consumed (m^3) is obtained from the analysis of the hydrologic balance, and real evapotranspiration is calculated from measurements using the technique of surface renewal. Rainfall data were measured using a rain gauge installed at the site. Irrigation water was measured and applied using a gauged watering bucket. In the article "Evaluating water productivity of tomato, pepper and Swiss chard under clay pot and furrow irrigation technologies in semi-arid areas of northern Ethiopia" more detail about the agronomic data was presented [13]. A comparative study has been undertaken between bar shaped clay pot and furrow irrigation on tomato, pepper and Swiss chard plots in Mekelle University Campus, Tigray, Ethiopia. Plant height for both tomato and pepper was measured every week using a ruler starting from 30 days of transplanting until maturity. The number of fruits per plant and yield were measured during the cropping season, and the results showed that there were five successive harvests of tomato and Swiss chard whereas there were only two harvests from the pepper crop.

2.5. Economic evaluation: comparison based on a cost/benefit relation (CBR)

An analysis of cost/benefit relation (CBR) was done dividing the present value of the total benefit by the present value of the total cost for each farm; the larger the resulting index, the more efficient is the project. In general, a larger CBR indicates that the project is economically viable, and this also indicates that the technology used is economically efficient.

2.6. Statistical analysis of field data

Analysis of variance was conducted using a statistical program to evaluate the efficiency of water use, biomass production, crop yield, and plant height, width, and fruits, among other variables. Implementation of demonstration units of the Africa partnership was conducted with more than 60 farmers and 12 Agricultural Agents trained and provided with training manuals in the local language (**Figure 1**).

Figure 1. Images showing the demonstration of the project in Africa. Source: Araya and Africa team.

The interactive training and demonstrations delivered to university students, farmers and extension agents have contributed to enhancement of knowledge of using clay pot technology, which has contributed to enhance food productivity in dryland areas of Ethiopia. Cooperation in scientific knowledge sharing and development of partnerships with Brazilian Embrapa scientists has also been enhanced.

2.7. Amazon/Brazil: demonstration area

In Brazil, the process was fully automated and the experiments were installed in the community of Lavras, city of Santarém, in a fruit garden using agroforestry. This change in strategy led to substantial gains to the project. The project leadership in Brazil actively worked to automate the low-cost irrigation process. The entire pottery process was documented with video and photographs while the artisans from Icoaraci Center made the clay pots, and equipment for monitoring of parameters such as soil and air temperature and relative air humidity were purchased. The irrigation apparatus, comprising pipes, floats, connectors, hygrometers, gutters, and water tanks were purchased and the whole process in Brazil was automated. A Demonstration Unit (DU) was installed at Embrapa Eastern Amazon, in Belém. During the 39th Agriculture and Livestock Fair in Santarém, a lecture was given to demonstrate the low-cost technology for efficient water use (**Figure 2**).

http://dx.doi.org/10.5772/intechopen.80605

Figure 2. Images showing the experiments installed in the Amazon sharing knowledge obtained from the Brazil partnership. Source: Martorano and Brazil team.

3. Results and discussion

The water seeps out through the micro pores of the clay pots with relatively slow flow and larger surface wetting time—area coverage around the roots of plants. Contrarily, perforated clay pots leak the water much quicker through the macro- and micro-pores and have relatively shorter wetting time—area coverage ratio. On the other hand, the difference between perforated bar and round types is simply the shape of the pots which affects the area coverage along the rows of the Swiss chard plant. Round types were not as suitable as the bar type (of the same capacity) for rows of Swiss chard crops due to their wetting area coverage along the two sides of the bar. Therefore, among the tested clay pot designs, the bar shaped perforated clay pot designs were evaluated as best in term of biomass and economic water use efficiency. The economic and biomass water use efficiency for the perforated bar clay pot design was higher than that of the bucket irrigation system. The other advantages with perforated bar clay pots over the bucket type is that the water source is inside the soil thus evaporation is almost nil and there is also less probability of occurrence of leaf disease due to watering and this ultimately improves the biomass and water use efficiency.

The highest economic performance was obtained in furrow irrigation during the first harvest due to higher investment in clay pots. However, after analyzing return on investment (six consecutive harvests) the bar shaped clay pot irrigation was highly superior in economic performance compared to the furrow irrigation practices. A marginal rate of return indicated that 478.18, 258.82 and 221.47% was obtained in Swiss chard, pepper and tomato, respectively. Sensitivity analysis also indicated that adoption of the findings is feasible and practical.

Comparisons of the irrigation methods using the clay pots showed that there was a significant difference ($p < 0.05$) for tomatoes irrigated in rows. The authors [13] infer that water availability was adequate and uniform in the clay pots compared to the irregular availability in the soil for the row-irrigated crops. The cumulative yield of the three vegetable crops irrigated using clay pots were significantly superior ($p < 0.05$) than crops irrigated in rows. There was a 30% increase in yield in the system where water was replaced using clay pots compared to the system wherein green pepper was irrigated in rows. These results show that the technology that irrigates using clay pots can be used even in conditions that use brackish or salty water. Furthermore, the yield of tomato was 32% greater using clay pots than for those irrigated in rows, thus confirming the efficiency of this system. Similarly, the yield of Swiss chard showed an increase in biomass of 51% using clay pots. This increase can be explained by the fact that Swiss chard has a shallow rooting system which facilitates the absorption of water in this system using clay pots as compared to irrigation in rows [4].

3.1. Expected results and importance of IrrigaPot technology: Amazon/Brazil

The expected results consist of the development of an irrigation technology using different intelligent solutions for the replenishment of soil water for different types of crops in Brazil (**Figure 3**), an increase in the number of farmers that understand the principles of this technology and that are able to develop new alternatives for the fabrication and use of these clay pots, development and testing of different forms of clay pots capable of satisfying crop water demand at low cost to producers, use of the successful results to create a specific line of credit for farmers to be able to fully use the IrrigaPot user-friendly, low-cost technology for water replenishment in agricultural systems, an increase in the supply and diversity of agricultural produce during dry periods, and an improvement in the quality of life of small-scale farmers thus allowing them to remain in the rural area using low-cost technology and reducing losses due to seasonal drought. Furthermore, we expect an increase in food security in a situation where rainfall variability threatens the food supply, and that rainfall storage will guarantee a water source with low or no loss of rainfall collected in the rainy season.

This technology will also provide greater opportunity and time for formal education due to the reduction in labor necessary to irrigate crops in the dry season, and thus help to eliminate child labor that is common in areas that have streams and small rivers used to supply water to crops, and will provide new opportunities to women that, now with more free time, can dedicate themselves to other artisanal activities or pursue a formal education, and production costs will therefore be reduced due to lower labor demand. Additionally, estimates of evapotranspiration will be used to help to plan a cropping strategy that uses water efficiently, with the water footprint as an indicator of crop sustainability for crops that adopt the IrrigaPot technology.

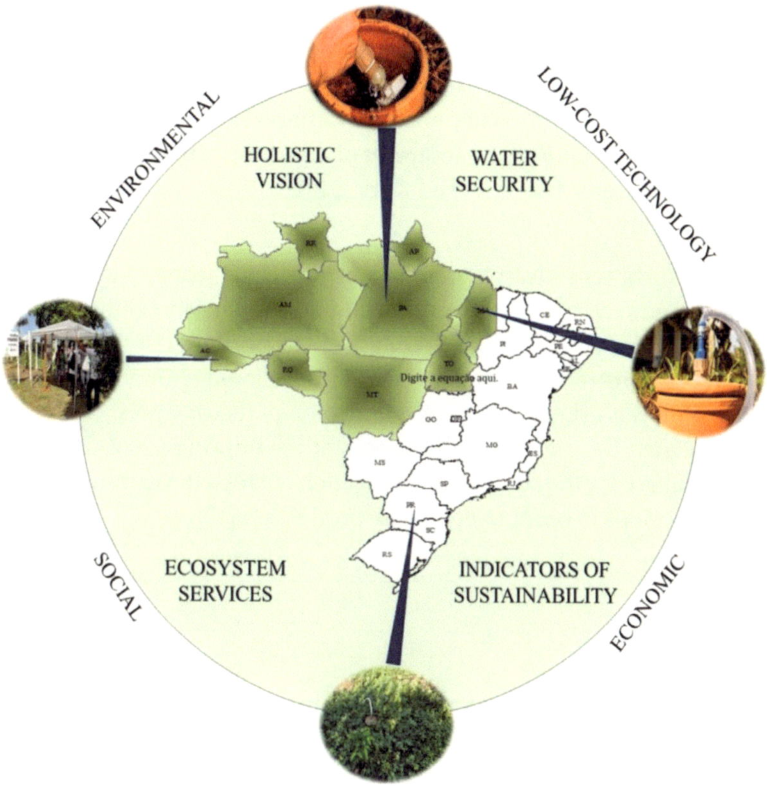

Figure 3. Monitoring and evaluation of water consumption, growth, development and productivity of crops (demands in northeast and southern Brazil).

The results from the Demonstration Units (DU) show gains in yield in the production of crops that use IrrigaPot technology by guaranteeing production during months with low rainfall. Agricultural producers express contentment due to the economic gains with the sale of products cultivated in areas using clay pots. In an interview [14] about the IrrigaPot Project, rural producers that cultivate using agroecological principles emphasized that the greater availability of water for plants guarantees gains and reduces preoccupation with crop failure, which allows them to engage in other activities on their property. The use of rainfall allows for the planting of species that previously could not be planted such as peanuts, tomatoes, peppers and achocha, associated with diverse fruit species. At the DU in the Lavras community (Santarém, Pará) plants maintained their production during the period of low rainfall in the region. Acerola plants and orange and tangerine trees guaranteed the availability of fruits at the local outdoor markets. Results published in different media sources [14–16] increased interest for the installation of this technology in new areas in many Brazilian States such as Acre, Tocantins, Amapá and Paraná (**Figure 3**).

The Brazilian Agricultural Research Corporation (Embrapa Amazônia Oriental/NAPT Médio Amazonas) presented this technology to organic producers, extension workers, and university professors and students highlighting the results from the Lavras community

(Santarém, Pará). The participants manifested interest in installing new units of the IrrigaPot project because they learned how rainfall could be used in agriculture during the dry months from August to November. The clay pots are maintained in the soil with 20 l of water and are able to meet plant water needs. It is important to emphasize that the use of rainfall waters reduced the blue water footprint of agriculture, because all the water used in irrigation comes from water that is stored during high-rainfall months in areas that adopt the IrrigaPot technology.

Using this process of collective learning, seminars were given wherein the fundamental techniques of irrigation with clay pots were presented in the DUs. Among the 150 participants in these activities, which were conducted in Altônia, in the northeast of the State of Paraná, the majority of them demonstrated interest in using this technology in their properties, but the largest barrier to this was established as the fabrication of the clay pots. This collective learning activity heightened the awareness of farmers of the importance of the adoption of this technology as a strategy for the sustainable cultivation of crops through replacement of water to soil during dry periods in order to guarantee production [17].

4. Conclusions

- The technology was found to be economically viable under conditions of small-scale growers, demonstrating the success of sharing scientific knowledge from the Brazil/Africa partnership.
- The sharing of knowledge about the IrrigaPot technology motivated agricultural producers from different regions of Brazil to adopt this technology as a strategy that promotes a pro-environment vision and that can bring social-cultural transformation with respect to the use of water resources in family-based agriculture.
- The partnership between the Brazilian Agricultural Research Corporation (Embrapa Amazônia Oriental), and the University of Makelle, da Ethiopia/Africa, with the incentive of the Program for Agricultural Innovation MarketPlace, integrated diverse specialists from Brazilian, African, Latin American, and Caribbean institutions in order to promote research projects and agricultural innovation and yielded successful results with the IrrigaPot project.
- This technology has gained the attention and interest of agricultural producers in different regions of Brazil and in neighboring countries, and also in Africa. These producers have installed the IrrigaPot system in different scales in diverse arrangements such as agroforests, and fruit and vegetable gardens, and have even adapted the system by modifying the type of clay pots.
- Society in general has increased pressure on governments and regulating agencies in order to provide incentive to agricultural producers to produce using with less irrigation water with the goal of reducing the environmental footprint of agriculture, principally the blue water footprint, and the IrrigaPot technology represents a sustainable practice for the sustainable replacement of soil water.

Acknowledgements

The authors would like to thank the MKTPlace Project and CNPq (Brazilian National Research Council) for financial support.

Author details

Lucieta G. Martorano[1]*, Araya A. Berhe[2], José Reinaldo da Silva Cabral de Moraes[3], Ayllan Rayanne da Silva Lima[4], Douglas Cavalcante Costa[3,5], Aline Michelle da Silva Barbosa[3] and Marcelo Coelho Marques[3]

*Address all correspondence to: lucieta.martorano@embrapa.br

1 Embrapa Eastern Amazon/NAPT, Santarem, Brazil

2 Mekelle University Campus, Tigray, Ethiopia

3 School of Agricultural and Veterinarian Sciences, São Paulo State University (Unesp), Jaboticabal, Brazil

4 Federal University of Amazonas, Brazil

5 Environmental Engineer, University of Pará State, Brazil

References

[1] Daka AE. Chapter 7 Clay pot sub-surface irrigation as water-saving technology for small-farmer irrigation in Development of a technological package for sustainable use of Dambos by small-scale farmers [PhD thesis]. South Africa: University of Pretoria; 2001. Available from: http://upetd.up.ac.za

[2] Okalebo JA, Home PG, Lenga FK. Pitcher irrigation: A new irrigation technique to curb the effects of salinization. In: Proceedings of the 7th Conference of the Society of Agricultural Engineers on Engineering the Economy. Nairobi, Kenya: Jomo Kenyatta University of Agriculture and Technology; 1995. pp. 15-21

[3] Bainbridge D. Buried clay-pot irrigation: A little-known but very efficient traditional method of irrigation. Agricultural Water Management. 2001;**48**(2):79-88. Available from: www.sciencedirect.com

[4] Araya AB, Martorano LG, Girma A, Habtu S, Kebede H, Hadgu KM. Comparative efficiency evaluation of different clay pots versus bucket irrigation system under Swiss chard (Beta vulgaris subsp. cicla) growers condition in Northern Ethiopia. Malaysian Journal of Medical and Biological Research. 2014;**1**(3):122-127

[5] Tsegasy Wolde-Georgis. Testing the Use of Clay Pots Sub-surface Irrigation Methods for Dry Land Farming in Atebes, Ethiopia, Progress Report to the Directors of Conservation,

Food & Health Foundation. Consortium for Capacity Building, Boulder: Univ. of Colorado; 2010

[6] Martorano LG, Bergamaschi H, Dalmago GA, Faria RT, Mielniczuk J, Comiran F. Indicadores da condição hídrica do solo com soja em plantio direto e preparo convencional. Revista Brasileira de Engenharia Agrícola e Ambiental. 2009;**13**:397-405

[7] Martorano LG, Bergamaschi H, Faria RT, Dalmago GA. Decision strategies for soil water estimations in soybean crops subjected to no-tillage and conventional systems, in Brazil. In: Techopen: Problems, Perspectives and Challenges of Agricultural Water Management. 1. Ed. 2012. United Kingdom. Available from: http://cdn.intechopen.com/pdfs-wm/31515.pdf

[8] Snyder RL, Spano D, Paw U KT. Surface renewal analysis for sensible and latent heat flux density. Boundary-Layer Meteorology. 1996;**77**:249-266

[9] Paw U KT, Snyder RL, Spano D, Su HB. Surface renewal estimates of scalar exchange. In: Hatfield JL, Baker JM, editors. Micrometeorology in Agricultural Systems, ASA Monograph No. 47. Madison, Wis: ASA-CSSA-SSSA; 2005. pp. 455-483

[10] Spano D, Snyder RL, Duce P, Paw UKT. Surface renewal analysis for sensible heat flux density using structure functions. Agricultural and Forest Meteorology. 1997;**86**:259-271

[11] Allen RG, Pereira LS, Raes D, Smith M. Crop evapotranspiration—Guidelines for computing crop water requirements. In: FAO Irrigation and Drainage Paper 56. Rome, Italy; 1998

[12] Allen RG, Periera L, Howell TA, Jensen ME. Evapotranspiration information reporting: I. Factor governing measurement accuracy. Agricultural Water Management. 2011;**98**:899-920

[13] Gebru AA, Araya A, Habtu S, Wolde-Georgis T, Teka D, Martorano LG. Evaluating water productivity of tomato, pepper and Swiss chard under clay pot and furrow irrigation technologies in semi-arid areas of northern Ethiopia. International Journal of Water. 2018;**12**:54-65

[14] Available in: http://g1.globo.com/pa/santarem-regiao/vem-com-a-gente/videos/t/edicoes/v/projeto-irrigapote-e-esperanca-para-produtores-da-comunidade-lavras/6073649/.2017

[15] Available in: http://revistagloborural.globo.com/Noticias/Agricultura/noticia/2017/01/potes-de-argila-sao-usados-na-irrigacao.html. 2017.

[16] Availabe in: https://www.embrapa.br/en/busca-de-noticias/-/noticia/19749409/potes-de-argila-saoalternativa-de-irrigacao-de-baixo-custo-no-brasil-e-na-africa

[17] Siqueira AP da S, Martorano LG, Moraes JRSC, Siqueira TTS, Silva TMG, Grossi-Milani R.Irrigapote: Aprendizagem coletiva na utilização de tecnologia de irrigação sustentável. Revista Educação Ambiental em Ação. N.64, 2018, Brazil. Available from: http://www.revistaea.org/artigo.php?idartigo=3229

Chapter 6

The Humidity of the Volcanic Soils and Their Impact on the Processes of Mass Removal in Colombia

William Chavarriaga Montoya

Additional information is available at the end of the chapter

http://dx.doi.org/10.5772/intechopen.80399

Abstract

This chapter describes some aspects of the formation of soils derived from volcanic ash, especially soils classified according to the soil taxonomy as Andisols of the mountains of the central mountain range in Colombia, cultivated with pastures for the production of milk, meat, and potatoes. General erosion, caving, mudslides, and landslides reach and cover large urban and rural territories characterized by a high rainfall regime with decadal records of 227 days of rain per year and with total loss of arable, agricultural, and productive layers. This chapter summarizes aspects of the research carried out by the author in these soils, through the description of profiles on 52 pits, field and laboratory analysis of their physical and some chemical properties for the understanding of the moisture storage capacity, and explains the mechanisms that govern their physical properties, composition, and interaction between particles and fluids and, consequently, their intense erodability and high moisture-retention capacity as a detonating mechanism of the processes of erosion and mass removal of secular occurrence in the soils of the region with high population densities causing hundreds of deaths and incalculable economic losses.

Keywords: volcanic ash, mass removal, erodability, landslides, humidity

1. Introduction

An important part of the Colombian territory is located mainly around and in the vicinity of the volcanoes and is covered with deep mantles as deposits of volcanic ash soils modeling the landscape of mountains and especially the central mountain range.

Soils of residual origin evolve from in situ weathering, and normally, they are characterized by a finer granulometry close to the surface where the alteration has been more intense. Despite this generalization, there are residual soils that reflect greater alteration in depth; this is often the case of soils derived from volcanic ash [13].

The soils derived from volcanic ash are those formed from the weathering of deposits of materials from volcanic ejections. According to the Committee for the Recognition of Soils [24], these soils are called Andisols, a name derived from ando soil; etymologically, "an" means dark and "do" means soil in Japanese language [10, 21, 27].

The central concept of the Andisols covers two fundamental aspects: (1) parental material of volcanic origin (ash, pomace, slag, pyroclastic, etc.) and (2) soils whose colloidal fraction is dominated by non-crystalline materials.

Under this concept, the specific properties of these soils have been attributed basically to the predominance of allophane in the colloidal fraction; however, the results found by Shoji and Ono [22] in soils without the presence of this mineral showed that the properties of the Andisols are not necessarily given by the allophone and indicate that the Al-humus and Fe-humus complexes also influence the properties of these soils [10].

Based on these results, new criteria were established to define the Andisols as soils developed from volcanic ejections or volcanoclastic materials whose colloidal fraction is dominated by non-crystalline materials and/or Al-humus complexes. It was also determined that the andic properties are the result mainly of the presence of significant amounts of Al-humus, allophane, imogolite, or ferrihydrite complexes [10, 16].

The physical, mechanical, and chemical properties of these soils make them considered as being of great importance worldwide due to their high productive potential, high carbon and nitrogen accumulation, high storage capacity, and improved water quality [10, 23].

Around the volcanic zones of the entire American continent are deposits of residual soils formed from the weathering of volcanic ash. Studies on similar soils and their performance in engineering works in regions such as Indonesia, New Zealand, India, Dominica, and Japan show that this type of soil has unusual properties compared to sedimentary soils [18, 20, 28]. In: soils derived from volcanic ash in Colombia [13].

2. Localization and distribution

The soils that currently cover the regions surrounding the volcanoes of the Andes Mountains in Colombia have their origin in pyroclastic materials that emanated during the volcanic eruptions of the last 25,000 years [9]. These deposits correspond to residual soils formed from the physical and chemical alteration of volcanic ash. Worldwide, volcanic ash soils represent approximately 0.84% of soils and are located predominantly in tropical regions [10, 13, 17, 21].

The soils derived from volcanic ash in Colombia occupy about 11.6% of the national territory and are located in regions of significant demographic and economic growth. In the Colombian coffee zone, it is estimated that about 350,000 ha of soils grown with coffee correspond to Andisols. These soils extend from the Eje Cafetero (Departments of Antioquia, Caldas, Risaralda, and Quindío) in the center of the country, to the departments of Tolima, Cauca, and Nariño to its south west.

3. Formation of volcanic soils

Volcanic ash is generated from the fragmentation of magma and materials in the cone of the volcano from previous eruptions [2, 13, 29]. Three mechanisms have been identified as the main generators of volcanic ash: the rupture of the magma due to vesiculation, the fragmentation due to high thermal stresses, and the pulverization of the lava in the walls of the volcano's chimney during eruption.

The mechanism of ash formation defines the block or vesicular morphology. The block ashes have flat surfaces resulting from the vitreous fracture of the magma. Vesicular ashes may have water drop textures or surfaces formed by the rupture of the material through areas that had air bubbles [13, 29].

The amount of water consumed in the transfer of thermal energy into mechanical energy also affects the production of volcanic ash. Dry eruptions (completely consumed water) lead to the formation of thickly laminated lapilli layers and thick ash layers (scale: dm–m). Wet eruptions (partially consumed water) lead to thin ash layers (scale: cm) [2].

Volcanic ash is composed predominantly of light primary minerals and mainly volcanic glass [14]. This primary mineral plays an important role in the formation of the minerals currently found. In a more advanced stage of alteration of the volcanic glass, halloysite is formed, a quasi-argillaceous primary mineral that is less evolved as a gel with a 1:1 Si/Al ratio. Most of the ashes that have led to soil formation in Colombia are dacitic, rich in plagioclase feldspar, volcanic glass, amphiboles, and pyroxenes, and poor in quartz [1, 13].

Residual soils derived from volcanic ash are developed through processes of physical and chemical alteration of volcanic ash deposits (dissolution, leaching, and precipitation of compounds). These processes of alteration transform the minerals, the shape and size of the particles, and the porosity. Its influence is controlled by climatic conditions and weather. Climatic conditions (such as precipitation, temperature, humidity, and wind) determine the presence of available fluids for chemical reactions, the rate at which these reactions occur, the migration of compounds, and the erosion, among other processes [4, 26]. Time, on the other hand, governs the sequence for the synthesis of secondary minerals and the distribution of particle sizes.

As a soil-forming factor, the effect of the parent material is more important in the initial stages of soil formation than in advanced stages. The weathering of the parent material depends on the presence of acidic or basic minerals. In general, acid minerals (e.g., quartz, feldspar, hornblende, mica, etc.) are more resistant to weathering than basic minerals (e.g., olivine, pyroxene, and calcium plagioclase [13, 26]).

During weathering, an elemental composition rich in Si, Al, and base cations (e.g., Na and Ca) is generally obtained. The Si and the basic cations are dissolved and removed from the surface layers and the Al tends to remain. As the climate becomes more humid, greater dissolution occurs and more aluminum (Al) is removed [13, 14, 30]. The mechanisms of dissolution and leaching are very important for the formation of soils derived from volcanic ash since they lead to highly porous surface areas and the availability of the necessary solutions for the synthesis of secondary minerals.

4. Soil-water relationship

In a general way, it can be said that the structure, the state of efforts, and the flow of water in any type of soil change when it is exposed to the intense cycles of drying and wetting, typical of the climatic conditions of the tropics. These changes affect the physical properties and mechanical behavior of the soil, which can lead to geotechnical problems (e.g., erosion, slope instability, etc.).

Soils derived from volcanic ash in Colombia are located in regions where a bimodal rainfall regime occurs during April to May and October to November and very dry periods occur between these stages. During periods of low precipitation and high temperature, high water evaporation occurs between the pores of the soil, causing its drying.

The evaporation produces contraction and increase of the suction forces in fine soils (silts and clays), for the states of complete saturation or partial saturation, respectively. The desiccation evolves occasionally toward the formation of cracks. These cracks can be understood as a consequence of the stresses produced by desiccation. Cracks in the surface of the soil make up areas susceptible to problems of erosion and instability, often observed on slopes with little plant cover, continuously exposed to drying processes. On the other hand, during humid periods, characterized by permanent and intense rains, the infiltrated water reduces the capillary effects and causes volumetric changes that can lead to swelling or collapse of the soil structure [13].

5. Erodability

In Colombia, the natural slopes in soils of volcanic origin reach heights between 10 and 20 m with slopes greater than 60° [8, 13, 19]. Despite this, the slopes are susceptible to instability, erosion, and cracking depending on the climatic conditions and vegetation cover. In the Colombian Coffee Region, landslides detonated by intense rainfall or locally intense earthquakes are often reported. These landslides can have a high potential for destruction in densely populated areas in mountainous reliefs of great length and high slope.

The soils of the region are characterized by steep slopes of 30° (67%) to 35° (78%), extensive slope lengths; the shape of the concave slope is favorable to the accumulation of surface and sub-surface waters. In addition to the detonating agent, the occurrence of a landslide is determined by previous conditions related to deficient plant cover, or the misuse or management of the soil, the poor disposition of agricultural production systems, the indiscriminate felling of forests for planting of pastures and livestock production and their precarious management and essentially physical causes inherent or intrinsic to these soils.

The coffee axis is located in a tropical zone that presents great climatic changes due to altitude changes and has a bimodal climatic regime given by two humid periods and two dry periods. The zone receives an annual precipitation varying between 1500 and 2250 mm. Surface landslides (depth < 1.5 m) are usually activated during periods of heavy rains, April to May and October to November, in which the accumulated rainfall during 1 or 2 days exceeds 70 mm [13, 25].

The superficial soils predominant in the area have deficiencies in the properties of resistance to the cut, since they are recently formed volcanic ash, unconsolidated, and sandy (Ruiz and Cerro Bravo volcanic complex in the Department of Caldas). These materials generally have low plasticity and cohesion due to their loose grain condition with sandy textural appreciations. The cohesion is drastically reduced (or even disappears) when the soil becomes saturated (reduction in the suction capacity), during the occurrence of intense rainfall, for example (the suction is lost and the natural cements dissolve).

The landslides have a flat and irregularly shaped surface defined by the contact between the layer of soils derived from volcanic ash and the layer that underlies it, composed of materials of vulcano-detrital origin, that are moderately or slightly weathered and/or evolved and they often come in slices. Slides of greater depth (depth: 3–10 m) are produced with detonating precipitation less than 50 mm, when the previous accumulated precipitation exceeds 200 mm [13, 25]. Dramatic differences in the permeability of these strata layers or horizons of these soils lead to the formation of a hung phreatic level that reduces effective efforts and increases instability or susceptibility to erosion.

6. Causes and effects of masal removal

Erosive processes are due to natural causes such as contact between geological units, in particular, a geometrically unfavorable contact between the upper volcanic ash (sandy and permeable and without aggregation) and the underlying igneous and metamorphic sedimentary rocks (compact, massive, and impermeable). This contact coincides with the fault surface of many of the landslides that have occurred and favors the accumulation of water that infiltrates through permeable surface of volcanic ash.

High torrentiality of permanent and intermittent drainage channels and lines exists in the region. Trees and very heavy shrubs on the crown of steep slopes generate a significant overload and negative "lever action."

The deforestation of the protection areas of the micro-basins, and the areas dedicated to pastures in the study area, becomes an accelerating factor due to the lack of protection of vegetation cover that counteracts the runoff associated with degadation phenomena gives origin to loss of soils and biodiversity and the alteration of the hydrological cycles of the basins or rivers of the region. This determines that areas of productive vocation, which are close to the micro-basins, that have lost their protective capacity of the ecosystems of strategic interest, are also affected due to the factors that undermine the stability of the soil, thus diminishing the potential to offer environmental services, of which the populated communities of the region are beneficiaries, limiting the production processes, and, therefore, their social and economic life.

Other determining factors of the drastic hydrological imbalances of the micro-basins of the region, which contribute significantly to the increase of flows, both surface water and infiltrated, which are the cause of landslides and mass erosion phenomena, are as follows:

- Increase in the change of land use from forests to paddocks. It has produced a drastic hydrological imbalance of micro-basins, significantly increasing the flows of surface and infiltrated waters.
- Excavation at the base of slopes and their over steepness, during the road construction processes.
- Deficiencies in road rainwater management works (transverse, without debris to stable and/or well-protected sites, and without internal structures to dissipate energy). Specific fillings in some areas of the road corridor, with low technical specifications and coinciding with sites of subsidence and settlements.
- Deposit of the materials resulting from the road cut, on the adjacent slopes, without any type of confinement. These "hillside fillings" coincide with the failed soils of some recent landslides.
- Specific problems of inadequate catchment, conduction, and delivery of surface water served in local homes (lack of channels and downspouts, deliveries of sewer networks to the hillside, soft areas without waterproofing, etc.).

The deforestation of the areas of interest for the protection of the micro-basins and the presence and increase of the areas in natural pastures in the study area are some of the causes of the decrease in water flows, which are associated with degradation of soils and aquatic and terrestrial flora and fauna and the alteration of the hydrological cycles of the basins, when climatic variables reach the most critical levels. The productive areas to intervene surrounding the micro-basins as ecosystems of strategic interest are also affected by climate change, which affects soils, reducing the supply of this environmental service to the beneficiary communities.

7. Technical support

According to PLA [15], amorphous clays, high in allophane, are the main determinants of the very particular physical and mechanical properties of Andisols. They are responsible for the

http://dx.doi.org/10.5772/intechopen.80399

development of low-density bulk floors, high porosity, high water retention (high saturation, field capacity, and tension of 1.5 MPa), and high limits (upper plastic limit or liquid limit—LPS and liquid plastic lower limit—LPI) of plasticity. The retention of available water (field capacity humidity at 1.5 MPa) is also usually high and limit liquid or water flow in the form of water is near, in soils not altered to the point of saturation.

Although the gravimetric water retentions are usually very high (up to 2–3 times the mass of dry soil when saturated), they are not so much on a volumetric basis due to the low apparent densities, although they are still higher than in other soils. The high retention of moisture even at high voltages and the poor connection between pores means that in humid climates, even with good drainage, conditions of poor aeration at shallow depths that restrict root development remain in the Andisols. In any case, to achieve such high moisture retentions requires a degree of weathering of volcanic ash, with formation of halloysite and accumulation of organic matter, since with very recently formed ashes, generally with sandy loam to gravel, the volumetric capacity Water retention is usually very low [15].

With drying, up to 30–50% of the water-retention capacity and a large part of its plasticity are irreversibly lost. It has been pointed out that the change of the plasticity indexes with the drying of the soil is the main property that distinguishes the Andisols from other soils where crystalline clays predominate.

The drastic and irreversible changes of properties of the Andisols derived from changes in humidity have much to do with the erosion processes in these soils.

The greater the inclination of the soils, the instability increases soils and, with it, the greater the susceptibility to mass movements, the more rainwater is infiltrated and less lost by runoff (accumulation). Mass movements depend on the interaction of several factors, especially slope; lithology; soil type; intensity, duration, and continuity of rainfall; surface and internal drainage conditions; vegetation cover; and management.

By virtue of the above, it is technically demonstrated that in Andisols, where a limiting layer has been formed for internal drainage at shallow depths and a high rate of surface infiltration is maintained, increases in moisture content negatively affect the stability of the soil material facing landslides by:

1. Increase in pore water interstitial pressure, which reduces the flow resistance in saturated soil over the restricted drainage layer.

2. Development of a hydraulic gradient or pressure in the direction of flow below the surface that can gradually lead to sub-surface erosion.

3. Lubrication of the limiting layer or sliding plane, which facilitates the movement of the material above it.

4. Increase in the mass of moist soil, sometimes 2–3 times its dry mass.

5. Decrease in the cohesion between particles and aggregates and once the soil is saturated, development of positive pressures in the pores [15].

8. Studies and research

Chavarriaga (2014) studied and investigated the physical and chemical characterization of soil profiles. Reference: evaluation of causal factors, effects and feasible management alternatives, the problem of erosion and mass removal of soils in the Maltería—Las Margaritas road transect, right slope of the Chinchiná River, via Magdalena "Department of Caldas-Colombia."

The investigation was carried out to identify and diagnose the problem of soil erosion processes in the area of influence of the Maltería-Las Margaritas road transect via Magdalena, right slope of the Chinchiná River, to technically evaluate the factors involved and the causal relationships—intervening effect on the problem of erosion and mass removal of soils, weighing risks and impacts, investigated about the factors related to the technical nature of the problem of deterioration of the soil resource, and its alternatives for improvement or mitigation, of the general impacts and develop the physical-chemical knowledge of the problem of environmental deterioration of soils in the area of influence of the Maltería-Las Margaritas road transect via Magdalena, Municipality of Manizales; Secularly converted into a factor of great environmental and socioeconomic impacts, aggravated in the winter periods of the area, which lead to problems of large soil losses, landslides, road restrictions, and all kinds of risks, which compromise important resources of the region, as losses of landscape, biodiversity and human lives.

8.1. Methodology

For the purposes of sampling, the digital cartographic information provided by CORPOCALDAS (Autonomous Regional Corporation of the Department of Caldas), stratified in three altitudinal ranges: high, medium, and low, considered as representative of the study area, was taken as a basis. The type of sampling applied was of a random nature and was carried out using functions of the ArcGis program based on a number of four repetitions of each combination of the variables "coverage" and "altitudinal range," resulting in a total of 52 sampling points (52 pits, duly geo-referenced). The resulting systems are shown in **Table 1**. The soil samples were made by opening pits of 1 × 1 × 1.50 m and making samples in each of them by soil profile (2–4 samples per profile according to horizons and profile morphology), which were processed for analysis in terms of physical and chemical variables. By groups of pits according to their altitudinal position and vegetation cover (5 coverings), a format or spreadsheet for the description of soil profiles was prepared (52 profiles) taking into account the methodology of soil surveys described by Cortés and Malagón [3] and the FAO profile description guide [7], both references updated according to the description method of the Geographic Institute Agustín Codazzi (IGAC) [12]. Soil chemical analyses were carried out in the soil laboratory of the Caldas University and the analyses for the physical variables in the soil physics laboratory of the National University of Colombia, Palmira-Valle. Both the chemical and the physical information were processed by correlation analysis for their interpretation and mapping according to their geo-referencing.

Table 1 indicates the edaphic systems under evaluation with their respective coverage and altitude ranges. The information on soil cover are indicative and taken from CORPOCALDAS and verified in the field, were studied, sampled, and analyzed the soils by means of pits as

No.	Coverage	Altitude range
1	Weedy grass	High: >2800 m above sea level
2	Secondary vegetation	High: >2800 m above sea level
3	Secondary vegetation	Medium: 2600–2800 m above sea level
4	Secondary vegetation	Low: 2400–2600 m above sea level
5	Mosaic of pastures with natural spaces	High: >2800 m above sea level
6	Mosaic of pastures with natural spaces	Medium: 2600–2800 m above sea level
7	Mosaic of pastures with natural spaces	Under: 2400–2600 m above sea level
8	Dense forest of high ground	High: >2800 m above sea level
9	High, dense forest of firm ground	Medium: 2600–2800 m above sea level
10	Dense forest of the mainland	Low: 2400–2600 m above sea level
11	Clean grass	Height: >2800 m above sea level
12	Clean grass	Medium: 2600–2800 m above sea level
13	Clean grass	Low: 2400–2600 m above sea level

Table 1. List of systems under evaluation (coverage and altitude range).

stipulated by the international guides of description of soil profiles. The altitudinal information was suggested by researchers to facilitate its analysis.

The mosaic illustrates the different systems of coverage and their altitudinal position and allows to observe the little spatial variability of the soils, preserving similarities in their morphology and their genesis or their own genetic homogeneity or inheritance provided by the ancient deposits of pyroclastic volcanic materials. The ancient and recent volcanic events in a certain way have shaped the landscapes themselves where the profiles of exposed and supra-lying soils are located and studied to the lithological formations or litho-units dominated by igneous rock materials predominantly but with the participation of shales and other metamorphic materials. In general, this is the panorama of strata or horizons evidencing eminently volcanic features whenever an attempt has been made to discover the soil to such depths edaphologically speaking and that have enabled world literature to highlight the particularities of our soils known as volcanic or volcanic ash (volcanic ash soils).

On the other hand, the exposed mosaic allows a visual approach to obtain knowledge of reality in terms of the fragility of these edaphic ecosystems and therefore their immense susceptibility to erosion or mass removal and accompanying their physical attributes estimate in this study how are sandy and frank sandy textures, friable or loose consistencies, slightly plastic and slightly sticky, loose structures or those without structure in lower horizons markedly pyroclastic, not plastic and not sticky and without structure or loose consistency.

In this regard, the Geographical Institute Agustín Codazzi (IGAC) [12], in studies close to this research area concluded that the alternation of materials: ash-lapilli-pumice sands that have originated different horizons, A and C layers, show that a polycyclic development of these soils allow to deduce the different depositions of pyroclastic materials that have suffered degradation and reconstruct the history of their evolution; in effect, once the horizon was formed, it was buried by new materials, repeating in this way the different cycles of contributions of tephra or pyroclastic layers.

In any case, the presence of melanization, mineralization, humification, and structural development processes on the horizon indicates a pedogenic development slowed not only by the continuous rejuvenation of the materials but also by the very low temperatures.

The soils have originated from volcanic ashes alternating with sands, lapilli, and pumice. In superficial cases, well drained, they present several A horizons of dark colors with good structural development buried by volcanic sands that in turn are covered by lapilli and pumice; this indicates that they have suffered several periods of rejuvenation. In addition, the A horizon meets all the requirements of an umbric epipedon with andic properties, for which reason the soils have been considered as moderately evolved.

The physical-chemical dynamics of these soils is controlled by the presence of allophone, an amorphous material originating from the alteration of volcanic ash, constituted by Si in tetrahedral site, Al in tetrahedral and octahedral sites, and other octahedral ions with high variable load or high capacity of cationic exchange (CEC), 25–50 cmol(+)/kg of soil, anionic retention power (mainly phosphates), high affinity for humus and high porosity; and these allophones establish with it strong bonds that result in the accumulation of organic matter in the soil.

The humus-allophane interaction gives the soil particular properties such as high porosity constituted by many fine pores and medium observed in many cases and high retention of water or moisture at different tensions as a result of the high microporosity and the presence of allophane and organic matter.

The description and interpretation of the external and internal characteristics of one of the 13 modal profiles representative of the different coverages and uses and in accordance with the heights and their symbol are presented below. The methodology used follows the guidelines and procedures for description and interpretation of the 2013 IGAC in its semi-detailed study of Caldas soils.

8.1.1. External features of the SV24262 profile (2400–2600 m above sea level)

Taxonomy: Typic hapludand.

Cartographic unit: Cedral Consociation. Symbol: VS24262.

Geographical location: Department: Caldas. Municipality: Manizales Site: finca: El Cedral.

Geographical coordinates: X: 851051, 2843; Y: 1049817,0675; Height: 2444 m above sea level.

Landscape: mountain. Type of relief: Andean peaks.

Shape of the terrain: slopes, peaks, and troughs.

Lithology: alternating layers of volcanic ash, lapilli, and sands, on granitic lavas.

Environmental climate: cold and humid.

Average annual rainfall: 1800–2000 mm. Average annual temperature: 8–15°C.

Edaphic climate: temperature regime: mesic. Moisture regime: udic.

Erosion: Class: pluvial water. Type: furrows. Degree: moderate.

Mass movements: Class: deformations. Type: cow's foot. Frequency: frequent.

Surface stoniness: there is none.

Rocky outcrops: there is none.

Floods: there is none.

Encharcamientos: there is none.

Water level: not found.

Natural drainage: good (good).

Effective depth: moderately deep.

Limited by: alternating layers of pyroclasts.

Diagnostic horizons: Epipedon: umbric. Endopedon: there is none.

Diagnostic characteristics: andic properties, mesic temperature regime, and umbric epipedon.

Natural vegetation: secondary vegetation.

Current use: forest.

Limitations of use: cold weather and slope.

Described by: William Chavarriaga Montoya. Date: April 2014 (**Table 2**).

8.1.2. *Profile interpretation*

The soils of the Consociación el Cedral formed of volcanic ash are moderately deep, well drained, with moderate structural development. These soils have brownish and yellowish brown A/C genetic horizons respectively and umbric diagnostic horizon with andic properties, for which the consideration is reiterated as moderately evolved soils (**Tables 3** and **4**).

8.1.3. *Chemical characteristics*

The results of the chemical analyses indicate that they are strongly acidic reaction soils with pH values between 4.7 and 5.3 with restrictions for K and Mg, whose Potencial De Hidrógenos (pH) is extremely low; they have medium to high values of S and high values of matter organic of soil (MO) and N, with MO being responsible for the CIC due to the low presence of clays. They do not contain aluminum contents that represent a toxicity hazard for many plant species.

VS24262	Depth horizon (cm)	Main characteristics
	(0–5 cm) Oe	Mattress of live and dead roots
	(5–65 cm) A	A dark brown (7.5YR 3/2); sandy loam texture; structure: large, thin, and weak angular blocks; friable wet consistency; consistency in wet conditions: not plastic and not sticky; abundant macropores; abundant medium, thin, and thick roots; little macroorganism activity, positive reaction to NaF, pH 5.3; diffuse and wavy boundary
	(65–80 cm) C	Yellowish brown color (10YR 3/4); Tuff: without structure (loose)
	(80–105 cm) Bb	Dark brown color (10YR 4/3); sandy texture; fine granular structure; weak, loose, moist consistency; consistency in wet condition: not plastic and not sticky; abundant macropores; scanty roots; positive reaction to NaF, pH 4.7; clear and wavy boundary
	(105–X cm) C1	Dark olive brown (2.5Y 3/3); sandy texture; without structure; coarse grain not consistent

Table 2. Profile no: VS24262 internal profile features.

The physical-chemical dynamics of this soil is controlled by the presence of allophane, caused by the alteration of volcanic ash; this component has an affinity for humus, and establishes with it strong bonds that result in the accumulation of organic matter in the soil.

The humus-allophane interaction gives the soil particular properties such as high porosity, high water retention, and high capacity for nutrient retention (CICA); however, most of the electrical charge is not available to retain nutrients at the soil pH; this load only appears when the pH rises, such as occurs when the floors are limed. The load, which depends on the pH, is called variable load (CICV) and is the one that is present in this soil. A feature that distinguishes soils of volcanic origin, due to the presence of allophane, is the low availability of phosphorus; however, the analytical results of this soil show average contents of this element, undoubtedly due to the presence of apatite in volcanic materials.

8.1.4. Physical characteristics

The dark brown color of these soils is generated by the accumulation and high levels of MO in the first horizon resting on clear materials. The texture is sandy loam, while the laboratory

http://dx.doi.org/10.5772/intechopen.80399

Reference	Horizon	Altitude (m)	Bulk density (g/cm^3)	Real density (g/cm^3)	Total porosity (%)	Macro (%)	Meso (%)	Micro (%)	Dispersión coefficient (%)
SV24262-A	A	2444	1.05	2.41	56.43	24.25	13.25	18.93	11.42
SV24262-Bb	Bb	2444	1.01	2.55	60.39	43.84	6.81	9.74	13.33

Reference	Horizon	Humidity retention							
		Saturation (%)	0.1 b	0.3 b	1 b	3b	5b	10 b	15 b
CG24262-A	A	85.30	69.95	34.12	33.48	31.04	27.91	23.34	20.07
PL24262-C	C	27.53	22.57	11.01	10.80	10.02	9.01	7.53	6.48

Reference	Horizon	Altitude (m)	Sieve (#10) (2 mm) (%)	Sieve (#20) (0.84 mm) (%)	Sieve (#35) (0.5 mm) (%)	Sieve (#60) (0.25 mm) (%)	Sieve (#>60) (<0.25 mm) (%)
SV24262-A	A	2444	87.8	3.48	2.88	1.48	4.36
SV24262-Bb	Bb	2444	92.8	1.52	3.36	1.24	1.08

Reference	Horizon	Altitude (m)	Stability index	DPM (mm)	State aggregation	Hydraulic conductivity (K) (mm/h)	Moisture of threat (m^3)
SV24262-A	A	2444	0.08	5.10	94.16	145.51	3,041,010
SV24262-Bb	Bb	2444	0.06	5.33	97.68	138.90	626,831

Table 3. Physical properties VS24262.

results obtained indicate coarser textures (sandy and sandy loam) due to difficulties that arise in the analysis by interference of the allophone. The structure is in subangular, thin, and moderately developed blocks. The porosity is high, (56.43 and 60.39%) for horizons A and Bb, respectively, with a large predominance of macro pores. The apparent density presents low values, normal for soils derived from volcanic ash. The conditions of aeration and drainage are good. The consistency is friable, not plastic, and not sticky in the described horizons.

8.1.5. Humidity retention

Figure 1 illustrates the moisture contents (%) and the soil moisture tension (Bars), information that indicates that as the soil tension increases, the moisture content decreases and what is related in this measure when the tension is zero (0) the ground is at the point of saturation.

Such soil water behavior is evident for both horizons: A and Bb. The humidity retention is high at different stresses as a probable result of the presence of allophane and high levels of organic matter.

Id. sample	Height (masl)	pH	Aluminum (cmol(+)/kg)	Nitrogen (%)	O.M cold weather (%)	Phosphorus (mg/kg)	Potassium (cmol(+)/kg)	Calcium (cmol(+)/kg)	Magnesium (cmol(+)/kg)	Sodium (cmol(+)/kg)	Sulfur (mg/kg)
SV24262-A	2444	5.3		0.59	16.46	22	0.05	1.43	0.14	0.319	16.67
SV24262-Bb	2444	4.7	0.2	0.54	14.36	104	0.06	0.93	0.09	0.235	8.32

Table 4. Chemical properties VS24262.

Figure 1. Mosaic of soil profiles in pits for different cover systems and at different altitude ranges in volcanic soils in Colombia (secondary vegetation; pasture mosaic with natural spaces; weeds; weedy grasses; clean pastures; and high, dense forest mainland).

8.1.5.1. Humidity threat

This measure is considered to be of great value in the study and in correspondence with the studied problematic as it is the mass removal of the soils; its meaning allows us to understand the capacity of these soils to retain water and to increase in volumetric and gravimetric terms the natural condition of the soil, that is, its volume of water content and its corresponding weight. The analysis starts from the consideration of the apparent density expressed in dry weight (1.05 and 1.01) and that allows to calculate the weight of a surface of soil (1 hectare). Determining its saturation point (80.35 and 41.38%) allows to quantify the water that can hold the soil in each horizon and that correspond to values of 3,041,010 and 626,831 m^3; this in sum equals 3,667,841 m^3 of water per hectare in the investigation of moisture threat.

By virtue of the above findings, water or humidity threat allows us to suggest the potentiality of moisture retention at the time of sampling and the determination of the saturation

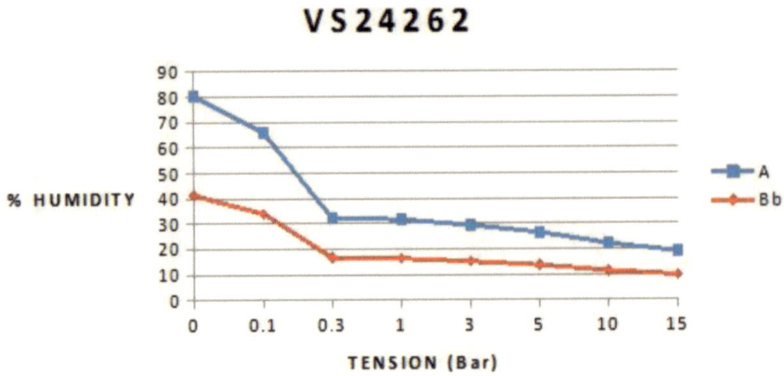

point for a soil that was not in rainy weather conditions. Such values constitute a powerful argument to estimate the extraordinary erosive capacity of the soil water storage and retention factor in the study and the increase of the soil susceptibility to the mass removals the pending factor is added to this, mountain relief, gravity, geomorphology, lack of protection of the soil of the natural or wooded vegetation cover and poor pasture management due to overgrazing.

In other words, the difference between the water retained to saturation and field capacity is the water that intervenes mainly in the phenomena of mass removals. Another implication is that: 3667.8 tons of water migrate from horizon A, toward C constituted by pyroclastics; there increases the speed of infiltration and in contact of moisture with the buried soil or horizon Bb the water hangs (drain hanging) thanks to the slope of the slope and the horizon becomes a plane of sliding.

The dispersion coefficients of 12.38% on average for the sampled horizons qualify the soil as stable. The variables DPM (weighted average diameter) with average value (5.22) as well as the state of aggregation (95.92%) allow one to assess the soil as very stable or in its defect state of very high aggregation (>90%). Conversely, the stability index < 1.0 warns of the presence of large aggregates that determine and indicate instability with aggregates greater than 5 mm, as confirmed by the DPM (5.22), and susceptibility to soil mass removals. Stability indexes greater than 1 would be ideal and would indicate predominance of intermediate aggregates well distributed in the soil. The usable humidity (12.8%) however for the soil is an average value of available water or useful water or vegetable water supply.

On the other hand, the saturated hydraulic conductivity (Ks) determined in the laboratory with values of 145.51 and 138.9 mm/h allows to identify the speed with which the water permeates the soil; therefore, it is a measure of the permeability as an intrinsic character of the soil. Such values indicate a very fast hydraulic conductivity and/or permeability.

9. Conclusions

The high humidity retention or high levels of saturation at the different tensions were confirmed as fundamental detonators of the mass removal of the studied soils, as a consequence of the instability of soil aggregates to water, high porosity, and high hydraulic conductivity and their relation with the mineralogy of these soils of volcanic origin, the high rainfall regimes of the region, the altitudinal position, slopes or inclination of the terrain that condition a high relative threat by mass movements.

Inter-variable correlations were found that facilitate explaining the phenomenon of mass removal in the area, among them some of significant order referring to the association between the variable "Humidity-Threat" and the organic matter for all the coverages analyzed. There is a negative effect of intervention on forests on the stability of soil aggregates.

There is a significant and positive relationship between the stability of the soil structure with the OM content and the degree of soil cover.

It is possible to explain the variations in the stability of aggregates, by the combined action of OM content and the degree of soil cover; however, this last variable is the most significant of the two.

The management of soils against mass removal should consult systems that involve minimal disturbance of the soil and the greatest possible protection through forest coverings, as ways to promote a stable structure and, consequently, promote the resistance of soils to water erosion.

Author details

William Chavarriaga Montoya

Address all correspondence to: william.chavarriaga@caldas.edu.co

Department of Rural and Natural Resources, Faculty of Agricultural Sciences, Caldas University, Manizales, Caldas, Colombia

References

[1] Arango JD. General characteristics and geotechnical behavior of volcanic ash in the Zona del Antiguo Caldas [master's thesis in Civil Engineering]. Bogotá, Colombia: University of the Andes; 1993

[2] Büttner R, Dellino P, Zinranowski B. Identifying magma-water interaction from the surface features of ashparticles. Nature. 1999;**401**(6754):688-690

[3] Cortés LA, Malagón CD. Soil surveys and their multidisciplinary applications, inter-American center for the integral development of water and land. Series: Soils and climate. Teaching material. SC-58, Mérida Venezuela; 1983

[4] Chadwick OA, Gavenda RT, Kelly EF, Ziegler K, Olson CG, Elliott WC, et al. The impact of climate on the biogeochemical functioning of volcanic soils. Chemical Geology. 2003;**202**: 195-223

[5] Chavarriaga MW, Gabriel Cruz C, Johan Cuervo C. Soil moisture retention and mass movement of volcanic soils from the "Sabinas" sector in Caldas, Colombia. Acta Agronómica Magazine, National University of Colombia Headquarters Palmira. 2017;**66**(4):588-597

[6] Tables O, Sisa R. Rainfall-landslide relationship in Manizales and its surroundings. Department of Civil Engineering, National University of Colombia; 2003

[7] Food Agriculture Organization (FAO). Manual of Description of Soil Profiles. Rome, Italy: Food Agriculture Organization; 1979

[8] Forero Dueñas C, Gálvez P, Fino, Ulloa. Studies of the structure of the volcanic ash of Armenia and its relationship with geotechnical behavior. In: X Geotechnical Conference of Colombian Engineering. Colombian Bulletin of Geotechnics. Colombian Geotechnical Society; 1999

[9] Herrera A, María C. Soils derived from volcanic ash in Colombia: Fundamental study and implications in engineering [doctoral thesis University of the Andes]. Bogota, Colombia; 2006

[10] Hincapié GE. Study and modeling of water movement in hillside volcanic soils. Palmira, Colombia: National University of Colombia. Faculty of Agricultural Sciences. Postgraduate School; 2011

[11] Agustín Codazzi Geographical Institute, IGAC. General Study of Soils and Land Zoning. Department of Caldas; 2004

[12] Geographical Institute Agustín Codazzi (IGAC). Semi-detailed soil study of the municipalities of Manizales, Chinchiná, Palestina, Neira, and Villamaría. Scale 1: 25,000. CORPOCALDAS. Department of Caldas; 2013

[13] Lizcano A, Herrera MC, Santamarina JC. Soils derived from volcanic ash in Colombia. International Magazine of Natural Disasters, Accidents and Civil Infrastructure. 2006;**6**(2): 167

[14] Nanzyo M. Unique properties of volcanic ash soils. Global Environmental Research. 2004; **6**(2):981-8555. Association of International Research Initiatives for Environmental Studies (AIRIES), Japan

[15] PLA SENTIS. Ildefonso. The erodability of the andisols in Latin America. Equatorial Soils. 1992;**22**(1):33-42. SCSS. Santa Fe de Bogota. Colombia

[16] Pochet G, van der Velde M, Vanclooster M, Delvaux B. Hydric properties of high charge, halloysitic clay soils from the tropical South Pacific region. Geoderma. 2007;**138**:96-109

[17] Quantin P. Properties and genesis of Andisols. Pédologue ORSTOM. 1986;**XXII**(1):70-74, 105-111

[18] Rao SM. Mechanistic approach to the shear strength behavior of allophonic soils. Engineering Geology. 1995;**40**:215-221

[19] Redondo GA. Geo-mechanical behavior of the volcanic materials applied to the slope stability of the Coffee Region [master thesis in Civil Engineering]. Bogotá, Colombia: National University of Colombia; 2003

[20] Rouse W. Volcanic soil properties in Dominica, West Indies. Engineering Geology. 1986;**23**: 1-28

[21] Shoji S, Nanzyo M, Dahlgren RA. Volcanic Ash Soils: Genesis, Properties and Utilization. Amsterdam: Elsevier Science; 1993

[22] Shoji S, Ono T. Physical and chemical properties and clay mineralogy of andosols from Kitakami, Japan. Soil Science. November 1978;**126**(5)

[23] Shoji S, Takahashi T. Environmental and agricultural significance of volcanic ash soils. Global Environmental Research Association of International Research Initiatives for Environmental Studies (AIRIES), Japan. 2004;**6**(2):113-135

[24] Soil Survey Staff. Soil Taxonomy. In: A Basic System of Soil Classification for Making and Interpreting Soil Surveys. 2nd ed. Washington, DC, USA: Agricultural Handbook 436, Natural Resources Conservation Service, USDA; 1999. p. 867

[25] Terlien MTJ. Hydrological landslide triggering in ash-covered slopes of Manizales (Colombia). Geomorphology. 1997;**20**:165-175

[26] Townsend FC. Geotechnical characteristics of residual soils. Journal of Geotechnical Engineering. 1985;**111**(1):77-94

[27] Wada K. The distinctive properties of andosols. Advances in Soil Science. 1985;**2**:174-229

[28] Wesley LD. Structural behavior of residual soils of the continually wetlands of Papua New Guinea—Discussion. Geotechnique. 1974;**23**:471-494

[29] Wohletz K, Krinsley D. Scanning electron microscopy of basaltic hydro magmatic ash. The Alamos National Laboratory Report, LA-UR 82–14; 1982. pp. 33-43

[30] Ziegler K, Hsieh JCC, Chadwick OA, Kelly EF, Hendricks DM, Savin SM. Halloysite as a kinetically controlled end product of arid-zone basalt weathering. Chemical Geology. 2003;**202**(3):461-478